Structural Integrity

Volume 34

The *Structural Integrity* book series is a high level academic and professional series publishing research on all areas of Structural Integrity. It promotes and expedites the dissemination of new research results and tutorial views in the structural integrity field. The Series publishes research monographs, professional books, handbooks, edited volumes and textbooks with worldwide distribution to engineers, researchers, educators, professionals and libraries. This series is managed by team members of the ESIS/TC12 technical committee. Topics of interested include but are not limited to:

- Structural integrity
- Structural durability
- Degradation and conservation of materials and structures
- Dynamic and seismic structural analysis
- Fatigue and fracture of materials and structures
- Risk analysis and safety of materials and structural mechanics
- Fracture Mechanics
- Damage mechanics
- Analytical and numerical simulation of materials and structures
- Computational mechanics
- Structural design methodology
- Experimental methods applied to structural integrity
- Multiaxial fatigue and complex loading effects of materials and structures
- Fatigue corrosion analysis
- Scale effects in the fatigue analysis of materials and structures
- Fatigue structural integrity
- Structural integrity in railway and highway systems
- Sustainable structural design
- Structural loads characterization
- Structural health monitoring
- Adhesives connections integrity
- Rock and soil structural integrity.

**** Indexing: The books of this series are submitted to Web of Science, Scopus, Google Scholar and Springerlink ****

Springer and the Series Editors welcome book ideas from authors. Potential authors who wish to submit a book proposal should contact Dr. Mayra Castro, Senior Editor, Springer (Heidelberg), e-mail: mayra.castro@springer.com

STRUCTURAL INTEGRITY AWARDS

STRUCTURAL INTEGRITY-SERIES AWARD

Regulations and Eligibility Criteria:

The Structural Integrity-Series Award is dedicated to the best contributions at Structural Integrity Book Series, as selected by the Awards Committee.

The best contributions are from books published in the Structural Integrity series in the form of research monographs, professional books, manuals, edited volumes, and textbooks. In cases where the awarded contributions have more than one author, a certificate will be sent to all authors of the book, book chapter, etc.

The award will be announced each year to the best contributions selected by the awards committee.

The awards committee is composed by the Editorial Advisory Board and Editors of the Structural Integrity - Series.

The winners of the Structural Integrity-Series Award have been:

2018: Jorge Luis González-Velázquez (author), Fractography and Failure Analysis. https://doi.org/10.1007/978-3-319-76651-5 (Book, Vol. 2)

2019: Mayer, A., Krasnikov, V., Pogorelko, V. (2019). Limit of Ultra-high Strain Rates in Plastic Response of Metals. In: Gdoutos, E. (eds) Proceedings of the First International Conference on Theoretical, Applied and Experimental Mechanics. ICTAEM 2018. Structural Integrity, vol 5. Springer, Cham. https://doi.org/10.1007/978-3-319-91989-8_60 (Book Chapter)

2020: Afanasieva, O.A., Zemskov, A.V. (2020). Unsteady Elastic-Diffusion Oscillations of a Simply Supported Kirchhoff Plate Under the Distributed Transverse Load Action. In: Gdoutos, E., Konsta-Gdoutos, M. (eds) Proceedings of the Third International Conference on Theoretical, Applied and Experimental Mechanics. ICTAEM 2020. Structural Integrity, vol 16. Springer, Cham. https://doi.org/10.1007/978-3-030-47883-4_34 (Book Chapter)

2021: Grzegorz Lesiuk, José A.F.O. Correia, Halyna V. Krechkovska, Grzegorz Pekalski, Abílio M.P. de Jesus, Oleksandra Student (authors), Degradation Theory of Long Term Operated Materials and Structures. https://doi.org/10.1007/978-3-030-43710-7 (Book, Vol. 15)

STRUCTURAL INTEGRITY AWARD OF MERIT

Regulations and Eligibility Criteria:

Every researcher can submit a proposal to the Awards Committee of the Structural Integrity – Series until November annually. The proposals should be submitted to Dr. Mayra Castro, Senior Editor, Springer (Heidelberg), e-mail: mayra.castro@springer.com

The proposal has to be seconded by another researcher, however, from a country other than that of the proposer (and should be from a different country than that of the candidate).

The proposal shall include a detailed justification of the candidate, usually covering one to two pages. It shall include a detailed CV of the candidate and a list of his/her five most important publications on the structural integrity area.

The Structural Integrity award of Merit will be announced each year.

The awards committee is composed by the Editorial Advisory Board and Editors of the Structural Integrity - Series, and two people outside the Book Series selected for just one year.

The winners of the Structural Integrity Award of Merit have been:

2018: Francesco Iacoviello (University of Cassino and Southern Lazio)
2019: Krishnaswamy Ravi-Chandar (University of Texas at Austin)
2020: Filippo Berto (Norwegian University of Science and Technology)
2021: available soon
2022: available soon

Grzegorz Lesiuk · José A. F. O. Correia ·
Shun-Peng Zhu · Hermes Carvalho ·
Kayode J. Olaleye
Editors

Composite Rebars for Reinforced Concrete Structures

Manufacturing, Testing and Applications

Springer

Editors
Grzegorz Lesiuk
Department of Mechanics
Materials Science and Engineering
Faculty of Mechanical Engineering
Wrocław University of Science
and Technology
Wroclaw, Poland

Shun-Peng Zhu
School of Mechanical and Electrical
Engineering
University of Electronic Science
and Technology of China
Chengdu, Sichuan, China

Kayode J. Olaleye
Department of Mechanics
Materials Science and Engineering
Faculty of Mechanical Engineering
Wroclaw University of Science
and Technology
Wrocław, Poland

José A. F. O. Correia
Department of Civil Engineering
CONSTRUCT
Faculty of Engineering
University of Porto
Porto, Portugal

Hermes Carvalho
Department of Structural Engineering
School of Engineering,
Federal University of Minas Gerais
Belo Horizonte, Minas Gerais, Brazil

ISSN 2522-560X ISSN 2522-5618 (electronic)
Structural Integrity
ISBN 978-3-031-96291-2 ISBN 978-3-031-96292-9 (eBook)
https://doi.org/10.1007/978-3-031-96292-9

This Springer imprint is published by the registered company Springer Nature Switzerland AG
The registered company address is: Gewerbestrasse 11, 6330 Cham, Switzerland

Preface

This book is organised into eight chapters dedicated to the topics: general overview of FRP rebars technology and applications, mechanical properties of FRP bars, damage prediction models, resistance to harsh environment, bond behaviour of FRP rebars embedded in concrete, flexural behaviour of concrete beams reinforced with FRP rebars, dynamic and cyclic performance of the FRP materials and, finally, a stat of the art on circular economy frameworks, composite materials and multiscale homogenization theories.

The first chapter provides an overview of FRP rebars technologies and applications. The civil engineering sector has driven the demand for materials that combine low cost, low density, environmental resistance, long durability and minimal maintenance. In this sense, the use of advanced composite materials in new construction and rehabilitation projects has consolidated itself in recent decades as a viable alternative. Glass fibre-reinforced polymer (GFRP) stands out as one of these composite materials, as it has been widely adopted in structural reinforcement techniques due to its advantageous properties. The pultrusion process has been the preferred manufacturing method, although it has undergone modifications in recent years due to the need to improve the adhesion of these bars to concrete. Therefore, Chap. 1 addresses several concepts related to this modified pultrusion process, highlighting its advantages and limitations. Additionally, this chapter also addresses the potential applications of this type of rebar, as well as its main advantages.

Chapter 2 is devoted to the mechanical properties of FRP bars, providing a design perspective, where the material's mechanical behaviour is crucial for selecting materials and geometries for specific design conditions. Therefore, this chapter presents experimental studies focused on analysing the mechanical performance of fibre-reinforced bars that reinforce concrete structures. This analysis is not limited exclusively to this type of application.

Chapter 3 addresses the damage prediction models to be considered in analysing and verifying FRP-reinforced elements. Several damage models have been used as criteria, including the von Mises criterion, the Tsai–Wu criterion (one of the most effective), the Hashin–Rotem criterion (considered the precursor to second-generation models), and others. Therefore, this chapter also addresses the application of these criteria to predict the failure of FRP (glass fibre-reinforced polymer) bars and discusses different approaches to modelling the post-failure behaviour of these materials. Finally, the potential use of these models in numerical simulations is explored, with a view to their application in advanced structural designs.

Chapter 4 focuses on the study of the resistance of this type of material to harsh environments. Reinforcing bars, in addition to mechanical stresses, are often exposed to aggressive environments, which can compromise their performance over time. This effect is especially relevant in offshore structures, where environmental conditions are particularly harsh. However, even in conventional applications, concrete can negatively influence reinforcement behaviour. While the degradation mechanisms in metallic materials are widely understood and documented, the same cannot be said for composite materials, whose behaviour in aggressive environments is more complex and still requires in-depth research. Therefore, this chapter presents and quantifies the results of an experimental investigation on the flexural behaviour of composite bars subjected to severe environmental conditions. Two different solutions were used to simulate these environments—alkaline water and salt water—with the tests conducted at elevated temperatures to accelerate the degradation effects.

Chapter 5 deals with the bond behaviour of glass fibre-reinforced concrete bars embedded in concrete. The structural performance of a concrete beam is influenced by several factors, such as the type of reinforcement used, its arrangement within the beam, and, most importantly, the bond behaviour between the concrete and the reinforcing bars. Among these aspects, this chapter focuses on analysing the bond between hybrid bars and concrete. This behaviour is crucial to ensuring the adequate performance of the structural element, since a poor bond can lead to premature failure of the beam, compromising its load-bearing capacity. Pullout tests were performed on composite bars produced by pultrusion to evaluate this phenomenon. This chapter compares the experimental results obtained with analytical models in the literature that describe bond behaviour. Good agreement was observed between the experimental data and the predictions of the theoretical models, thus validating their applicability.

Chapter 6 analyses and discusses the flexural behaviour of concrete beams reinforced with glass fibre-reinforced rebars. Glass fibre-reinforced polymer (FRP) rebars represent an effective solution for strengthening concrete structures, especially in aggressive environments. In addition to excellent corrosion resistance, these rebars offer advantages such as a high strength-to-weight ratio and electromagnetic neutrality, making them a promising alternative in various structural applications. However, for their use to go beyond the laboratory context and be adopted on a large scale in real projects, it is essential to conduct detailed investigations into the behaviour of these reinforcements under real service conditions. This chapter presents some of the results obtained by the authors from tests on reinforced concrete beams using hybrid FRP rebars, composed of a combination of glass and carbon fibres. Parameters such as displacements, deformations, cracking patterns and failure modes were analysed based on criteria established by North American and Canadian standards to evaluate the structural performance of the strengthened beams. The results demonstrated that the hybrid bar solution performed satisfactorily, highlighting its potential for practical application in concrete structures.

Chapter 7 evaluates the dynamic and cyclic performance of FRP materials. Although high-mass or high-velocity collisions are relatively rare, they can cause significant structural damage or collapse. These impacts can have a variety of origins, including wind-borne objects (wind 'missiles'), traffic accidents, or the use of weapons. Structural analysis and design in these scenarios are challenging due to the dynamic inertial loads involved and the impact of high deformation rates on material properties. Therefore, any advances related to impact resistance or the development of more accurate dynamic analysis and design methods are highly valued in structural engineering. For decades, experts have investigated the behaviour of plain and reinforced concrete under impact loads. This chapter, in turn, is dedicated to analysing the dynamic and cyclic performance of FRP composite materials, exploring their response under extreme loading conditions.

Finally, Chap. 8 presents an overview of circular economy frameworks, composite materials and multiscale homogenization theories. The world has faced serious environmental challenges and the depletion of natural resources, mainly due to past economic models and development processes. In this context, the circular economy has emerged as a promising strategy for promoting sustainable development. Several researchers and institutions have explored this model to increase resource efficiency, drive low-carbon strategies and reconcile industrial growth, environmental protection and human health. At the same time, composite materials have gained prominence in structural and civil engineering applications due to their advantageous properties, such as high mechanical strength, lightness, durability, corrosion and impact resistance, as well as good chemical performance and long service life. These characteristics also make composites attractive from a sustainability perspective, favouring their inclusion in circular economy strategies. With the growing use of these materials, the need for practical, accurate and functional analytical techniques for their evaluation arises. Among the methods commonly used in composites modelling are macromechanical and micromechanical analysis, each with specific limitations. Multiscale approaches have been widely adopted to overcome these limitations, allowing for integrating analyses at the microstructural and macrostructural scales. Therefore, this chapter presents an overview of the state of the art regarding the circular economy and the recyclable composites that can be used within this model. Furthermore, it seeks to introduce the main homogenization theories and multiscale approaches necessary to evaluate the behaviour of these materials efficiently and sustainably.

With this book, the editors and authors aim to contribute to the discussion surrounding the use of composite materials, including recycled composites, for engineering applications, such as civil engineering, mechanical engineering and offshore engineering, among others, where monotonic and cyclic mechanical behaviour, impact and environmental exposure, among others, are explored and analysed.

The authors would also like to express their gratitude to all the institutions that have financially (and not only) contributed to the book's preparation, i.e.:

- Research funded by the National Centre for Research and Development of Poland (NCBiR) grant number LIDER/40/0219/L-10/18/NCBR/2019;
- Wrocław University of Science and Technology, Faculty of Mechanical Engineering, Department of Mechanics, Materials Science and Engineering;
- University of Porto, Faculty of Engineering—This research was supported by UID/04708 of the CONSTRUCT—*Instituto de I&D em Estruturas e Construções*—funded by *Fundação para a Ciência e a Tecnologia*, I.P./MCTES through the national funds;
- University of Electronic Science and Technology of China, School of Mechanical and Electrical Engineering;
- Federal University of Minas Gerais, School of Engineering, Department of Structural Engineering.

The editors and authors of this book are very grateful to all those who directly and/or indirectly contributed to this document. Special thanks to the Springer team for all the effort and work expended in this book's preparation and graphic presentation.

Porto, Portugal — José A. F. O. Correia
Wrocław, Poland — Grzegorz Lesiuk
Chengdu, China — Shun-Peng Zhu
Belo Horizonte, Brazil — Hermes Carvalho
Wrocław, Poland — Kayode J. Olaleye
July 2025

Contents

About the Editors

Prof. Grzegorz Lesiuk has been employed at the Wrocław University of Technology (Faculty of Mechanical Engineering) since 2009. He obtained his doctoral degree in technical sciences, mechanics) in 2013, and habilitation (mechanical engineering) in 2020. His research activities are on fracture and fatigue mechanics of materials and structures, new materials and modelling of their durability under cyclic and static loads. He has published over 200 scientific papers in national and international scientific and technical journals, more than 100 conference papers, and one patent in the field of mechanical engineering/materials science. Grzegorz Lesiuk is the chairman of the Polish Group of Fracture Mechanics. He is also a member of the Polish Society of Theoretical and Applied Mechanics (PTMTS, a deputy member of the central authorities) and ESIS (European Structural Integrity Society). He is a co-author of an academic mechanics textbook and a monograph (Springer) on degradation problems of materials and structures. He is also the supervisor and co-supervisor of nine doctoral dissertations in Poland, Portugal, and Brazil. Professor Grzegorz Lesiuk is a co-organiser of many international and national scientific events, such as the 20th ICMFM (20th International Colloquium on Mechanical Fatigue of Metals), the 18th and 19th KKMP (National Conference on Fracture Mechanics), and the 1st VCMF (Virtual Conference on Fatigue of Metals). He was repeatedly awarded for his scientific activities: Membership of the Academy of Young Scientists and Artists operating at the President of the City of Wrocław (Poland)—nomination received from the Mayor of the City of Wrocław (2018), Rector's Award of the Wrocław University of Technology (2018, 2020), International ESIS Merit Award—TC12—2019 and international TOP 1% Reviewers (Publons and; Web of Science) awards—2019, Lierati Award (2019)—Emerald Journals Group. He is also the author of numerous expert papers and reviews. More than 95% reviews are performed for JCR journals—number of reviews: 600.

José A. F. O. Correia was born in 1984 in Peso da Régua, Portugal, and is currently a Professor and a Researcher at the CONSTRUCT institute, Faculty of Engineering, University of Porto (FEUP, Portugal). Since March 2018, He has been a Guest Professor at the Delft University of Technology (Netherlands). Also, He was an

Invited Professor at the University of Porto (2018–2019) and at the University of Coimbra (2016/09–2021/08), and He was an Associate Professor by invitation at UTAD (2022/03–2023/02). In 2007, José A. F. O. Correia obtained a Graduation in Civil Engineering (five academic years) from UTAD. In 2009, He received the M.Sc. in Civil Engineering (Structural Mechanics) from UTAD. He has a Specialization in Steel and Composite Construction from the University of Coimbra (2010). In 2015, He got a Ph.D. Degree in Civil Engineering from the University of Porto. In 2024, He obtained the Habilitation in Civil Engineering—Fatigue and Structural Integrity from the University of Porto. Since 2019, Doctor José A. F. O. Correia has been included in the world's top 2% scientists list of the Stanford University ranking. José A. F. O. Correia is co-author of more than 234 journal papers in the most relevant scientific journals, 200 conference proceedings, 18 book chapters, and co-author of 22 books (15 completed; seven ongoing). The candidate also edited more than 21 issues of scientific journals as a Guest Editor. He was/is a supervisor of doctoral and master's students in the areas of Civil and Mechanical Engineering—24 Ph.D. (11 completed; four to be completed soon—2025; nine ongoing) and 52 M.Sc.. (48 finished and four ongoing). José A. F. O. Correia received several distinctions and awards—by the Wrocław University of Science and Technology (Poland), University of Electronic Science and Technology of China (China) and Beihang University (China), FEUP Scientific Recognition Award 2019, 2020, 2021, 2022 and 2023, Highly Cited Paper (Clarivate)—2021 and 2022, Manson-Coffin IGF 2021 medal, Philo Mechanicus award 2023, IAAM Scientist Medal 2024.

Shun-Peng Zhu is a Professor in Mechanical Engineering at University of Electronic Science and Technology of China (UESTC), China. He received his Ph.D. degree in Mechanical Engineering from UESTC in 2011, and worked as PIF Fellow of Politecnico di Milano, Italy from 2016 to 2018 and research associate at University of Maryland, United States in 2010. His research which has been published in scholarly journals and edited volumes, over 150 peer-reviewed journal papers, explores the aspects: fatigue reliability; probabilistic Physics of Failure modeling; multi-physics damage modeling and life predictions. He received Second class Natural Science Research Awards of Sichuan Province (2022) and Chinese Society of Aeronautics and Astronautics (2022) and Chinese Society of Theoretical and Applied Mechanics (2025), First Prize of Technological Invention of China Association of Inventions (2022), three Most Cited Paper Awards of FFEMS Journal (2018-2022), Robert Moskovic Award (2024) and Award of Merit (2019) of ESIS-TC12, Top 100 Most Influential International Academic Paper Award of China (2018), Polimi International Fellowship (2016), Hiwin Doctoral Dissertation Award (2012), Best Paper Awards of 8 international conferences. He is featured as Elsevier Most Cited Chinese Researchers in the field of Safety, Risk, Reliability and Quality since 2018 and serves as Co-Editor-in-Chief of International Journal of Structural Integrity, editorial board member of several international journals like EFA, IJFatigue, FFEMS and book series.

Hermes Carvalho is a Steel and Composite Structures Professor at Federal University of Minas Gerais (UFMG)—Department of Structural Engineering, Brazil. The most important aspects of the Hermes Carvalho curriculum are the following: disciplines taken—Steel and Composite Structures; Structural Dynamics; Steel and Composite Bridges; supervisions in progress—five doctorates and three masters (Experimental and Numerical analyses of structures); supervisions finished—17 master dissertations; topics of research—Dynamic and Fatigue Assessment of Bridges and Steel Structures; Evaluation, Rehabilitation and Reinforcement of Steel Structures; Dynamic Assessment of Offshore Structures, Transmissions Lines and Composite Floors; New solutions for Composite Beams; Stability of Steel Elements; Structural Health Monitoring; coordinator of Structural Dynamics and Acoustics Lab; ex-coordinator and Professor of Post-Graduation Program in Structural Engineering at UFMG; member of the technical committee responsible for elaborating Brazilian codes related to steel buildings and steel and composite bridges (ABNT—Brazilian Association of Technical Standards); member of the board of three doctorates and 19 master's degrees; member of the board of seven thesis projects; 68 papers published in recognized journals, and 89 papers in Congress and Conferences.

Kayode J. Olaleye is a Ph.D. student at the Department of Mechanics, Materials, and Biomedical Engineering, Wroclaw University of Science and Technology, Wrocław, Poland. He is an author of about 18 publications, Scopus: h-index three and 20 citations, according to Google Scholar Citation, h-index three and 26 citations. His interests include, but are not limited to: Ballistic impact, nonlinear dynamics of materials, mechanical behaviour of materials, numerical modelling, FEM, ballistic materials, composites, and many others. He is currently investigating the analysis and influence of composite backing on impulse loads reinforced with the top layer of a spatial structure.

Chapter 1
General Overview of FRP Rebars Technology and Applications

Paweł Zielonka, Szymon Duda, Krzysztof Towarnicki, and Wojciech Błażejewski

Abstract To survive the unwelcome severe loading and hostile environmental conditions, the competition in the civil engineering markets typically enforces low-cost, low-density, and environmentally resistant materials with minimal maintenance and long service life qualities. Because of this, the use of sophisticated composite materials for new construction and renovation applications has improved to an acceptable level during the past few decades. Due to the characteristics of FRP, powerful strengthening techniques for various structural applications have become the main focus of many studies in recent years. To ensure price competitiveness with other industrial solutions, pultrusion was chosen as a manufacturing process. Pultrusion technology provides a high production efficiency with bars, whose main strength parameters are located along the axis of the profile. Evolution of the approach of composite bars to increase adhesion parameters to the concrete obtaining pseudo-ductility effects forced modifications to the standard production process. Several concepts of the modified pultrusion process with pros and cons will be discussed in this chapter, such as the benefits of the pull-winding manufacturing system. The FRPs approach has been effectively applied to improve underground infrastructures, silos, tanks, tunnels, buildings, and bridges. Due to their benefits, such as being lightweight, fatigue resistant, having high tensile strength, anticorrosion, and thermal insulation, FRPs have been studied as high-performance materials. This chapter will include a comprehensive description of the potential application of FRP rebars as well as its main advantages.

Keywords Fiber-composites · FRP bars · Unidirectional composites · Pultrusion technology

---

P. Zielonka (✉) · S. Duda · K. Towarnicki · W. Błażejewski
Faculty of Mechanical Engineering, Wroclaw University of Science and Technology, Wroclaw, Poland
e-mail: pawel.zielonka@pwr.edu.pl

G. Lesiuk et al. (eds.), *Composite Rebars for Reinforced Concrete Structures*, Structural Integrity 34, https://doi.org/10.1007/978-3-031-96292-9_1

1.1 Introduction

The applications of composite bar in the construction market are continuously increasing. Rebars, which are the main type of reinforcement used are glass fibers, are used mainly as a counterpart of steel bars due to their higher relative ultimate strength, resistance to chemical attack, harsh environmental corrosion, electromagnetic neutrality, and long-term strength [1, 2]. The wide advantages of glass fiber reinforced polymer (GFRP) provide enormous potential of application in industrial structures, off-shore construction, or other special sites. The specific of the applied loading conditions and strength requirements for the rebars defined the pultrusion process with several modifications as the best manufacturing technology. The pultrusion is an automated, continuous process dedicated to manufacturing profiles with constant cross section. This is a highly productive method that provides a high fiber content structure and reinforcement directed along the axis of the profile. The first applied pultrusion process was at the beginning of the 1950s, after that it was dynamically developed. The number of manufacturers increased from 20 companies at the beginning of technology to mostly 300 pultruders in the world in 2006. Most manufacturing machines are based on thermosetting resins. The pultrusion process begins with pulling the continuous rovings of fibers or mats from a series of creels through guiding systems into the resin bath. The reinforcements are impregnated with heated resin and guided to the heated die through the system, which shapes the fibers close to the final profile. After entering the heated die, the curing of the thermosetting resin is initiated by heat. The cured profile leaves the mold and comes to the section responsible for constant movement of the bars. At the end, the rod is cut into the required lengths using a cut-off saw without interrupting the continuous pultrusion process [3]. The requirements for the composite rebars enforce the modification of the pultrusion process. One of the key parameters, which forces changes in the method, is necessary to develop the outer surface of the rod to increase the adhesion to the concrete [4]. The modified surface is achieved using the axial strand of the fiber, which wraps around the rod and burnishes the structure of the bar [5].

In the following sections, various technological aspects such as material for a matrix, reinforcement, and additives used are mentioned. The impact on the rebar properties caused by differences between the manufacturing techniques and additional operations is explained. The process parameters are precisely notified. The commercial applications of the FRP rebars are presented in the last section.

1.2 Manufacturing Technology of FRP Rebars

1.2.1 Materials Used in FRP Rebars

The composite structure is a combination of reinforcement, matrix, and additives, whose whole composition is responsible for the resultant properties. Each component needs to be properly selected to another to achieve the desired effect considering such aspects as adjusted coating of the fibers for type of the matrix, selected matrix, and fibers for required environmental conditions.

Reinforcement

In recent times, non-metallic fibers are extensively used in construction sites for concrete reinforcement. The materials are being investigated for resistance to electrochemical corrosion [6], mechanical performance [7], and mechanical behavior of FRP reinforced concrete structures [8] to increase the applications of bars. Due to their specific properties, FRP bars could be widely used for bridge reinforcement [9], offshore structures to reduce maintenance cost caused by corrosion of steel reinforcement. The typical types of fibers are presented in Fig. 1.1. The most dominant and versatile material used for FRP bars is E-glass fiber.

The most widely used form of material in the pultrusion process to manufacture rebars for concrete reinforcement is roving. Fibers in the form of mats, textile could also be applied.

Matrix

The matrix is a component of the composite structure that is responsible for distributing the load between particular reinforcements. The mechanical parameters of the material and properties of the interface between substrates are crucial for the overall mechanical behavior. Thermosetting resins are common polymers used in

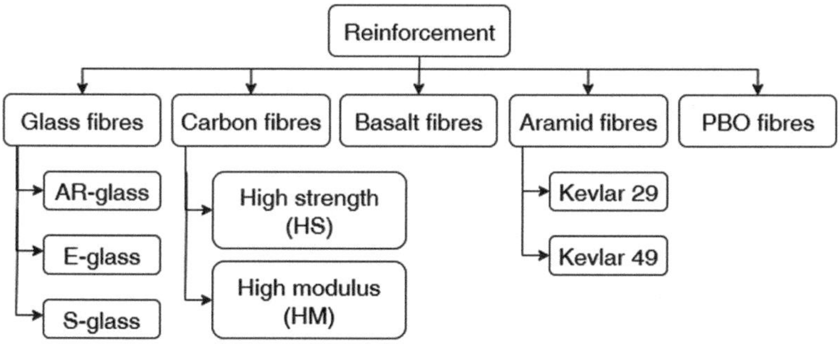

Fig. 1.1 Typical types of materials used as reinforcements in fiber reinforced polymer (FRP) rebars

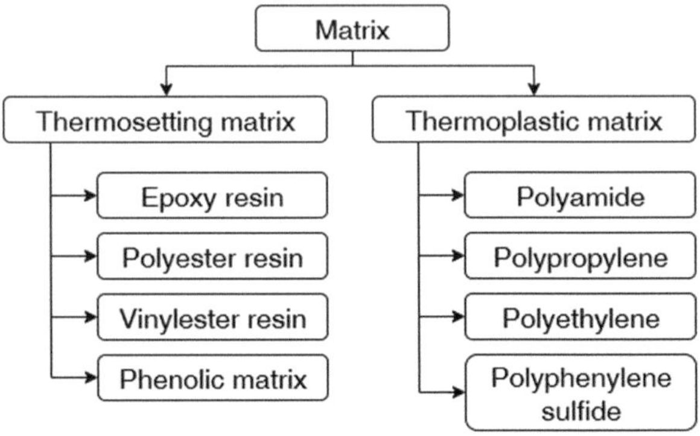

Fig. 1.2 Materials used for the matrix in pultrusion technology

pultrusion technology. Nowadays thermoplastic polymers are adjusted to manufacture rebars due to advantages such as bendability [10], and recycling. The materials commonly used in the pultrusion process are presented in Fig. 1.2 [11].

Thermosetting Resins

Epoxy resin is the most common material used for construction applications due to its good mechanical properties, strong stability, heat resistance, and low shrinkage. The polyester resins are chosen in the fields where reduction of costs is a more crucial aspect, and the principal mechanical parameters are sufficient for application. Another matrix system is selected for specific applications. Thermosetting resins are applied in the process mostly in the liquid form, where the low viscosity and curing time at higher temperatures are important parameters responsible for the proper process. Several methods used pre-impregnated roving during manufacturing.

Thermoplastic Matrix

The thermoplastic matrix has different technological aspects compared to thermosetting resins. The main difference is the form and viscosity of the matrix. Thermoplastic materials are often used in the form of filament or powder which is distributed on the surface of the reinforcement. During the application of heat, the matrix is melted and the fibers are impregnated. Typically, the viscosity of the thermoplastic material is higher than that of the thermosetting resin [12] (Fig. 1.3).

Additives

Additives are one of the ingredients of the composite structure, which improves the performance of the final product. The addition of the filler could adjust the material to specify environmental conditions, decrease the production cost, or improve mechanical properties.

Flame Retardants

Flame retardants are an important filler, whose main aim is to limit flame spread, reduce burn-through rates, and ensure minimal smoke generation and toxicity. In

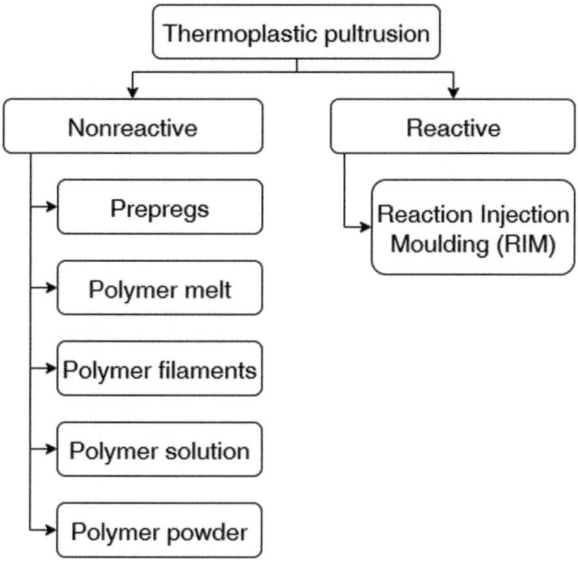

Fig. 1.3 The various aspects of thermoplastic pultrusion technology

rebars used for civil engineering applications, providing properties which ensure human safety are crucial aspects. In the pultrusion process, halogen-free additives such as aluminum trihydroxide (ATH), ammonium polyphosphate (APP), and triphenyl phosphate are commonly used [13–15].

Shrinkage Decreases Additives

Extreme temperatures of the die and rapid cure often lead to phenomena such as shrinkage of the resin. This effect provides for the loss of dimension, which could negatively affect the appearance of the components. Manufacturers reduce the amount of shrinkage using aluminum powder, talc, or thermoplastic polymers dissolved in styrene. One of the popular low-profile additives for unsaturated polyester resin is poly(vinyl acetate) PVA(c) [16].

Additives Improved the Toughness of the Material

The crucial aspect of the components used is their useful life. The material, which could have high static tensile strength, could be destroyed by low-level routine stresses in a short period of time. Rubber and related elastomeric products are mostly used to improve the flexibility of FRP components. The phosphorus-containing polysulfone additive could serve as a flame retardant and improve crack resistivity simultaneously [17, 18].

Release Agents

The crucial issue in composite manufacturing is to remove the cured components from the die. To deal with the problem, the release agents are added to the matrix. The usual Internal Mold Release (IMR) agents are waxes, fatty acids, or metallic stearates. The mechanism of interfacial layer formation between the mold and the composite could be due to the migration of IMRs to the outer layers or the change in solubility of the release agent between the cured and uncured matrix. The way of processing depends on various factors, such as resin type, chemical composition of the release agent, processing parameters, chemistry of the mold surface [19].

1.2.2 Hybridization Approach for Composite Rebars

Most composite bar products are manufactured using unidirectional glass fiber reinforcement with helical winding due to the cost and technological aspects. The main disadvantage of this solution is the brittle behavior of the structure and Young's modulus, which lowers the steel bars. Several approaches to the hybrid structure are proposed to eliminate the shortcomings of the solution. One of the proposed concepts is to use the wrapping layer with a different lay-up configuration, which could be applied using filament winding technology or braiding modules during the pultrusion process [20, 21]. Another approach is to use the different types of reinforcement in a single structure [22, 23]. One of the most important outcomes of the hybridization method is to achieve the pseudoductility effect, whose theoretical stress–strain curve is presented in Fig. 1.4. Pseudoductile behavior is a result of failure of the individual layer of the structure.

Another reason for the hybrid layout application is to adapt the structure to the prevailing environmental conditions [24]. Various types of reinforcement with specified additives could be applied to the bar to extend the useful life of the structure and ensure the best use of the material. The scheme of the hybrid sample solution is shown in Fig. 1.5.

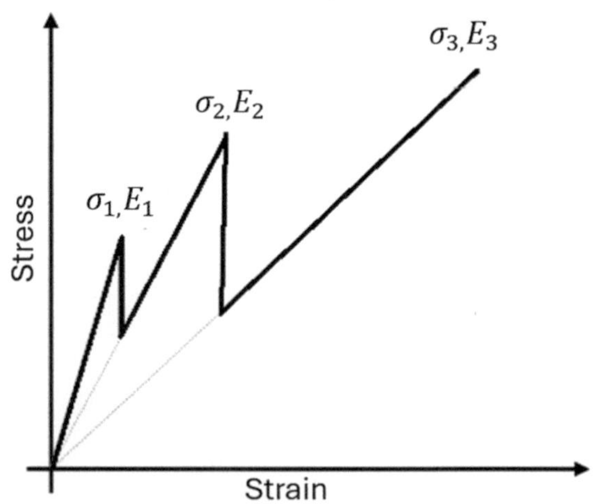

Fig. 1.4 Theoretical stress–strain curve of typical hybrid reinforcement bar [20]

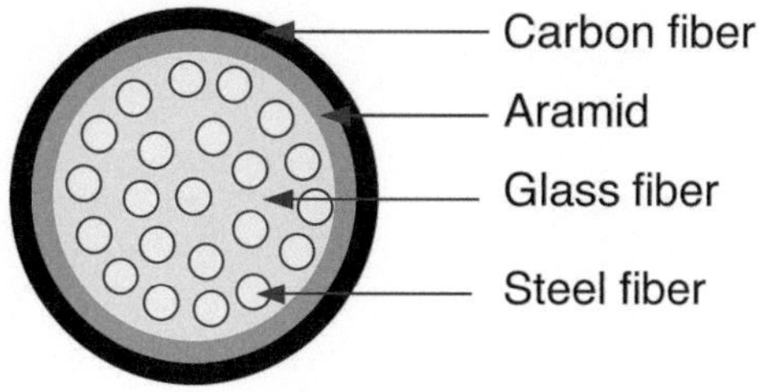

Fig. 1.5 The concept of hybrid bar with various types of reinforcement [23]

1.2.3 Comparison Between Rebars Manufacturing Technology

This section focuses on the differences between several composite rebar manufacturing technologies. Pultrusion is a continuous process, which is the main advantage, allowing reduction of the cost of the product. To accomplish the expectations for rebars, modification of technology needs to be done. The modification will be compared in a few sections with highlighted pros and cons. The comparison will be concentrated on the rebars based on the thermosetting epoxy system.

Standard Pultrusion Process

The outcome of the standard pultrusion process will be a reference for comparing the results obtained from the modified technology. The scheme of the process is described in Fig. 1.6. The bobbins are bobbed through the guide table to the impregnation system. In most applications, the external impregnation of the fibers by the matrix is the first step of the process. In several cases, direct injection to the mold is conducted, mainly due to the resin requirements and the reduction of volatile compounds in the production hall. The more specific description between the various concepts of fiber impregnation will be explained in the following section. The impregnated fibers are pulled through following eye guides, whose main rule is to perform the roving to demand cross-section to the heated die. In the mold cross-linking of the matrix will perform. The temperature of the heaters is selected to achieve a solidified profile at the end of the mold. Next, the rebars are cooled and the pulling system, responsible for constant movement of the profile, is obtained. The last operation is cutting a product for the specified dimension using saw.

The standard pultrusion process allows the bars to have a constant cross section with highly packed fibers. The microstructure of the rebars obtained is presented in Fig. 1.7.

The overall volume content of fibers in the profile could reach 65%, which is one of the highest in the composite manufacturing technique. Several resin-rich regions between strands of the fibers are noticed and could be a clue for better arrangement of the roving before heat die section. The pro of the method is the simplicity of the process, high tensile strength of the final rebar, and limitation of the internal defects such as voids and resin rich regions. The main disadvantage of this solution is the smooth surface of the profile, which has a direct impact on the results of the concrete

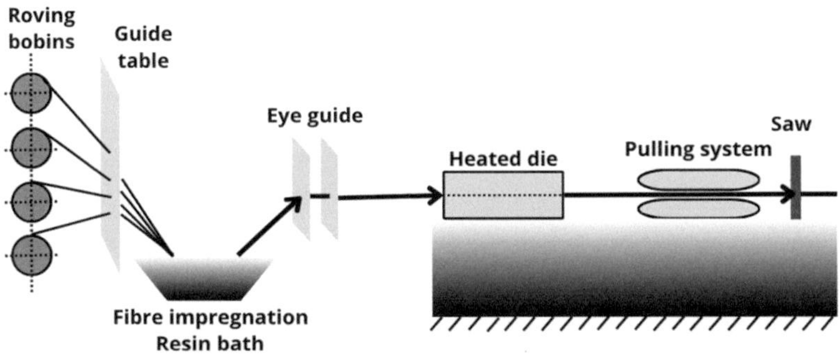

Fig. 1.6 Standard pultrusion process to manufacture composite profiles with a constant cross-section

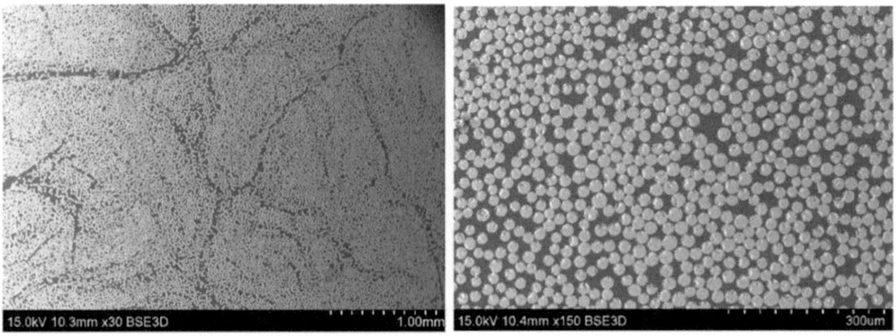

Fig. 1.7 Cross section of rebar manufacturing using a standard pultrusion process

pull-out tests. The surface could only be modified by sand coating or thread milling [4].

Pultrusion Process with Braiding Module and Heat Zone

Modifications of the pultrusion technology are conducted to achieve rebars with an extended surface. The machine was additionally equipped with a rotating disk module and heated zone as a replacement for the heated die shown in Fig. 1.8. The rotating disks are able to adjust the rotation speed to the axial velocity of the rebar to make the thread with the required angle. The bobbins with a guiding roving system are mounted on the disk. Radial fibers are impregnated using the resin from the axial roving. It simplifies the rotating disk equipment.

Mold replacement was carried out to maintain the shape of the barbed wire given by the rotating disk module. The heated zone could be equipped with various types of heaters such as thermal resistance heaters or using infrared and microwave radiation. The section is responsible for properly curing the matrix, and the application of the selected type of heater could have an impact on the final mechanical parameters of

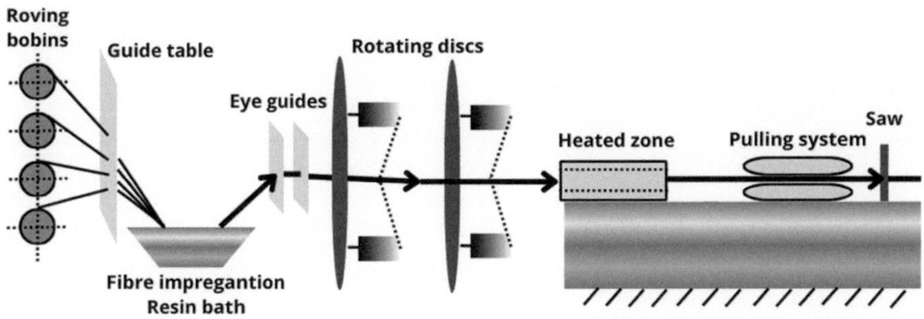

Fig. 1.8 Modified pultrusion process to manufacture threaded rebars for concrete reinforcement

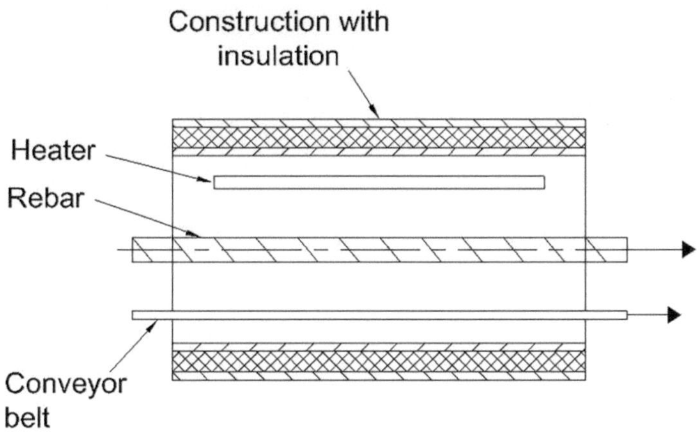

Fig. 1.9 Schematic of the heated zone with heaters

the product [25]. The heat zone scheme is described in Fig. 1.9. The conveyor belt is used to remove possible resin residues from the internal space of the heat zone.

The modification of the technology is to manufacture the rebars with extended external surfaces, which influence the adhesion properties of the concrete. The process ensures the high possibilities of rebar shaping as a result of the wide range of the axial and radial roving tension changes. The major disadvantage is the packing of the fibers in the rebars and the appearance of the void, which have a direct impact on the mechanical properties. The SEM images showing the microstructure of the bar are presented in Fig. 1.10. The weaker fibers packing could be observed to correspond to the figures from the standard pultrusion process.

Pultrusion Process with Braiding Module and Dual Heat Section

To overcome the difficulties in the previous section, the heated die was added to the technological line to consolidate the axial fibers and reduce the content of the resin in the rebar. As a result of the modification, the tension of the fibers increases due to the phenomena between the impregnated fibers and the heated die. The higher tension

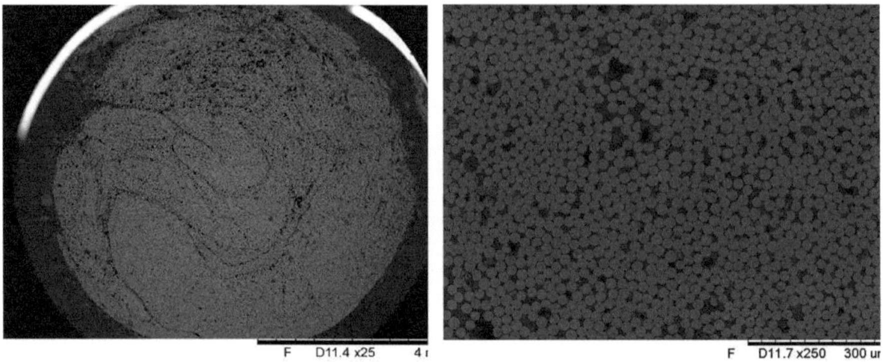

Fig. 1.10 Cross section of the bar manufactured using the modified pultrusion process with heat zone

of the core causes the smaller depth of the rebar that can be shaped. The scheme of the modified pultrusion process is described in Fig. 1.11. In this technology, resin is initially cured in the heated die but still in liquid form. The next step is to make the thread on the rotating disk section. The radial fibers are impregnated using resin from the axial roving. The interface between the winding and the core is weaker compared to previous braiding technology, due to the higher viscosity of the matrix. The final cross-linking of the bar is performed in the heat zone.

The microstructure of the bar is characterized by better packing and reduction of voids and resin in the structure, which is presented in Fig. 1.12. Modifications have an impact on the mechanical performance of the bar and better surface end. This method is a compromise between previous technologies.

Influence of Technological Parameters

Technological parameters are an important issue to obtain the component with required specification. The main parameters in the pultrusion process are the temperature and linear velocity. These and the other variables are described in the following section.

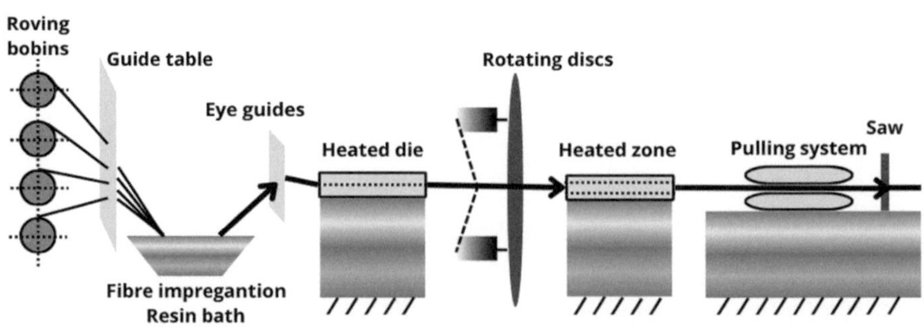

Fig. 1.11 Modified pultrusion process to manufacture threaded rebars for concrete reinforcement

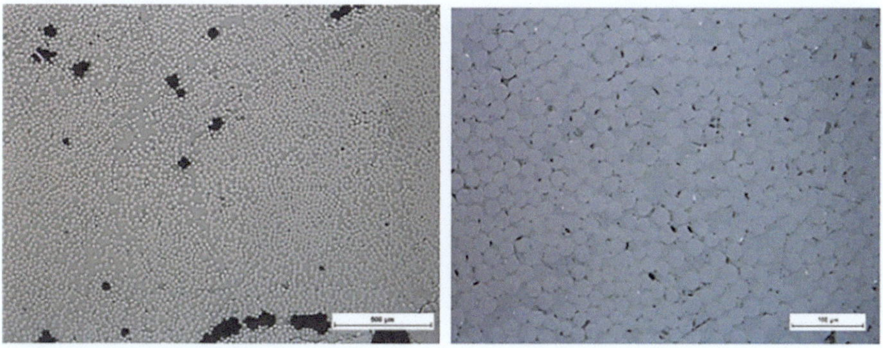

Fig. 1.12 Cross section of the bar manufactured using a modified pultrusion process with a heat die and braiding system (scale of 500 and 100 μm)

Temperature of the Heated Die/Heated Zone

Temperature is the first of two main technological parameters of the pultrusion process. Appropriate temperature settings result in obtaining the required mechanical parameters and should be adjusted for the specified matrix with additives and geometry of the profile. During the standard pultrusion process the following phenomena are presented in Fig. 1.13.

Manufacturing of rebars needs an additional setup due to the braiding module and non-constant cross section. The heat zone with various types of heaters needs to be applied for correct curing process. The scientist conducted research on the influence of the type of heaters, such as resistance, infrared, or microwave heat source, on the mechanical properties of the rebar [25–28]. The correct selection must be performed for the specified profile requirements.

Linear Velocity of the Pulled Bar

The linear velocity of the rebar corresponds directly to the temperature settings during the process. The increase in velocity requires higher temperature values to cure the profile in the specified section. After the speed increases for a certain level, the extent of the heating area is required to provide enough time for the curing process. Increased pressure inside the mold, caused by the increase in parameters, could suppress void formation and improve fiber wetness [29, 30].

Temperature of the Resin in Impregnation Systems

The temperature of the resin has an impact on the various parameters of the matrix. In thermosetting matrix systems, an increase in temperature causes a reduction of the viscosity and accelerates the curing process. In pultrusion technology, the temperature of the matrix is initially rising to improve impregnation abilities in the resin bath systems. The temperature depends on the matrix system used, typically oscillating between 60 and 80 °C.

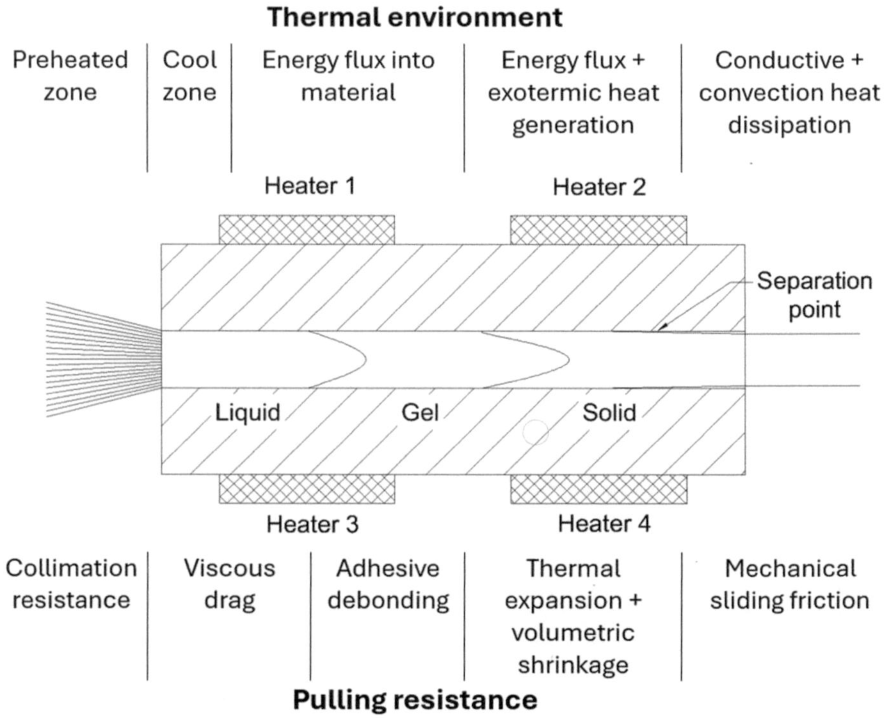

Fig. 1.13 Thermal environment and dissipation of pulling resistance in the heated die for the standard pultrusion process [3]

Tension of the Axial and Radial Fibers

The traction of the roving is a crucial parameter responsible for the proper shape of the bar. The ratio of the axial and radial fibers provides a way to achieve a specified depth of the thread, which could have an impact on the adhesion to the concrete or mechanical properties. The other parameter with influence on the geometry of the rebars is Tex of the roving. A higher Tex of the radial roving causes lower pressure of the fibers on the outer surface of the rebar and weaker compression of the outer axial fibers. The thread squeezing process on the bar surface is presented in Fig. 1.14. The geometry of the thread, such as the angle between the radial fibers and the axis, could be modified by changing the ratio of the linear velocity and the angular speed of the rotating disk.

The most difficult task is to predict the tension of the axial fibers. The tension of the roving is the summary of several factors such as tension on the bobbins, tension in the impregnation system, eye guide system, or heated die. Most of them depend on the geometry of the system and cannot be modified during the manufacturing process. The easiest way is to change the brake torque on the bobbins to adjust the tension to

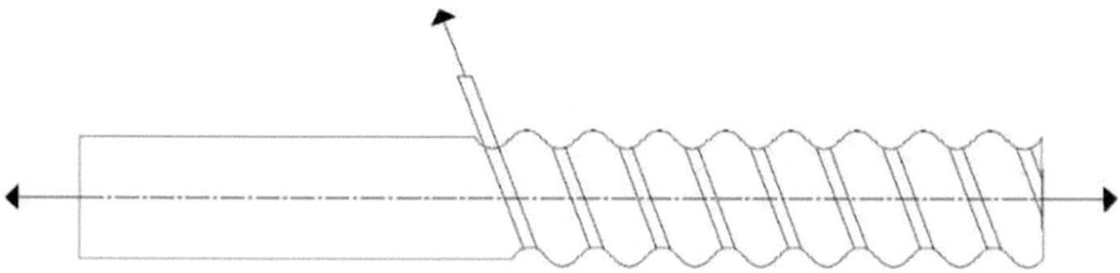

Fig. 1.14 The thread squeezing process on the rebar surface

optimize the process. The elementary components responsible for the tension of the axial fibers are highlighted in Fig. 1.15.

The tension generated in the guide table and impregnation system components could be reduced using minimum fiber bending angles. The small bending of the roving prevents the fibers from destroying during pulling. The generation of tension on the pin surface, which is the result of friction, is presented in Fig. 1.16.

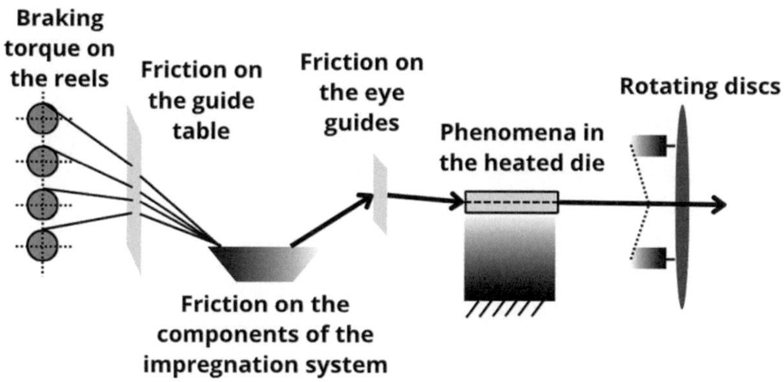

Fig. 1.15 Components of tension of the axial fibers

Fig. 1.16 Tension of the fibers generated on the pin surface

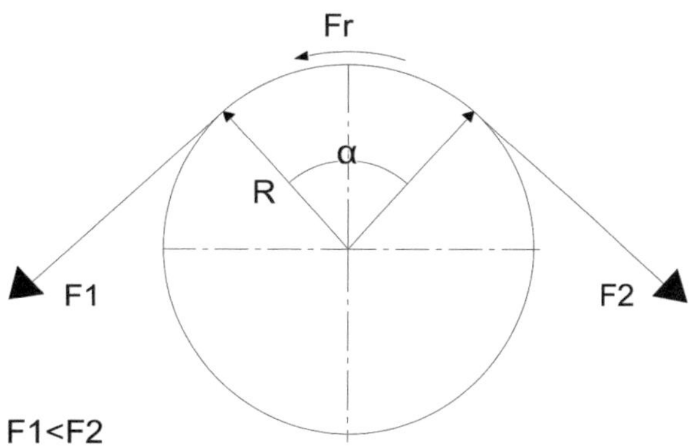

To ensure proper tension on the axial and radial roving, the braking system connected with the bobbins mountings is used. The brakes could work in several ways, such as using friction between the belt and the brake disk or electrical hysteresis. The working principle of the mechanical braking system is described in Fig. 1.17. The tension of the belt is controlled by the applied force to the spring. The spring regulation could be done manually or using servo motors. Another mechanism is based on the electromagnetic power brake system. After the application of electricity, the internal magnetic flux is created. The flux is transferred into a hysteresis disk that passes through the field. A magnetic drag on the hysteresis disk allows for a constant drag of the output shaft. In each type of solution, the important aspect is the changing value of the fiber tension, due to the decreasing radius of the reel during the process. The changes of the reinforcement tension must be compensated for maintaining constant parameters during technological process.

Tex of the radial roving

The Tex of the radial roving is one of the parameters responsible for the final geometry achieved for the rebars. Smaller values of the Tex leave the axial fibers with higher pressure in smaller areas. It is made to manufacture rebars with tiny thread. The pros of using smaller values of Tex are better impregnation of the radial reinforcement during winding. The systems not equipped with the roving impregnation system mounted on the rotating disk use the resin accumulated in axial reinforcement. During the winding of the radial fibers, the roving is impregnated. On a higher Tex, complications could appear with fiber impregnation or strength of the interface between radial and axial reinforcement.

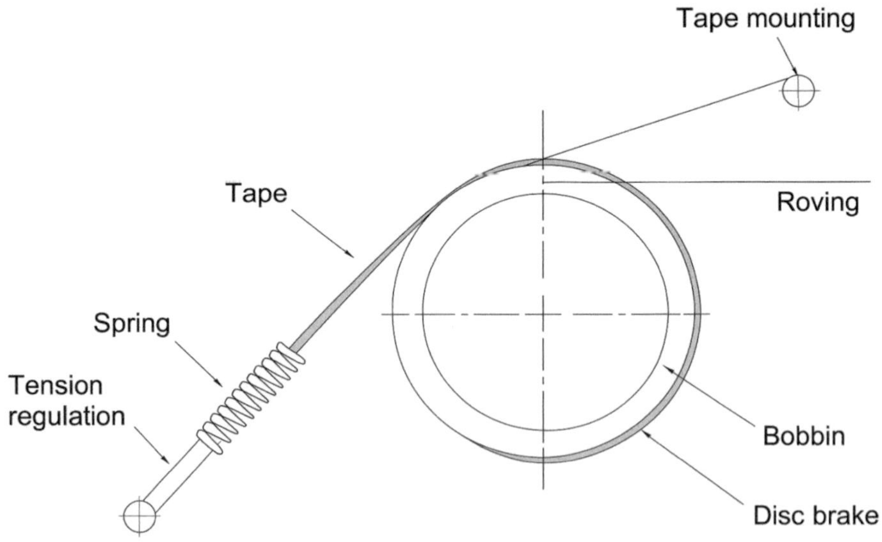

Fig. 1.17 Mechanical brake system

1.2.4 Rovings Impregnation System

The matrix system used in the composite materials is characterized by various parameters. To meet the resin requirements and ensure the best possible percolation of the fiber, several techniques of the impregnation system were developed presented in Fig. 1.18.

The most popular is the resin bath impregnation system, used in the two main variants: dip-type and drum-type. The dip-type is characterized by simple construction and the working method. The roving is anchored under the resin level and impregnated using the tension generated on the surface of the pins. The roving with excess resin is coming out of the bath and needs to be removed in the following operation. This kind of system could be easily used in pultrusion, where the resin content is strictly determined by the cross section of the die. The scheme of the system is presented in Fig. 1.19.

The second impregnation system, the drum-type, is used mostly in the filament winding machines because of the limitations of the resin in the fibers after the process. The resin is delivered by the rotating disk, which transports the material using surface adhesion. The thin film of the resin is used to impregnate the fibers by using pressure at the interface of the elements. The idea of the process is shown in Fig. 1.20.

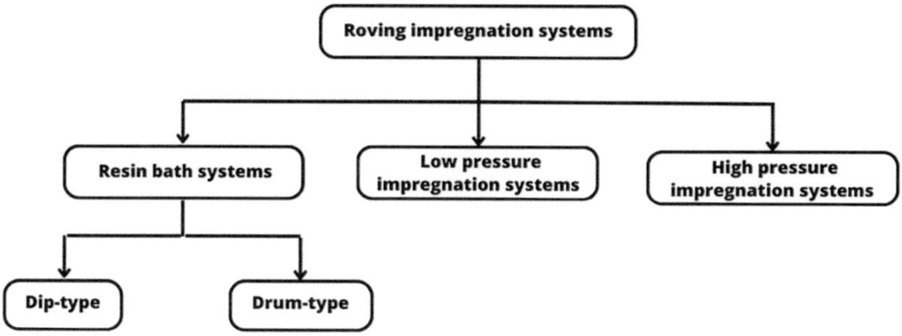

Fig. 1.18 Types of roving impregnation systems

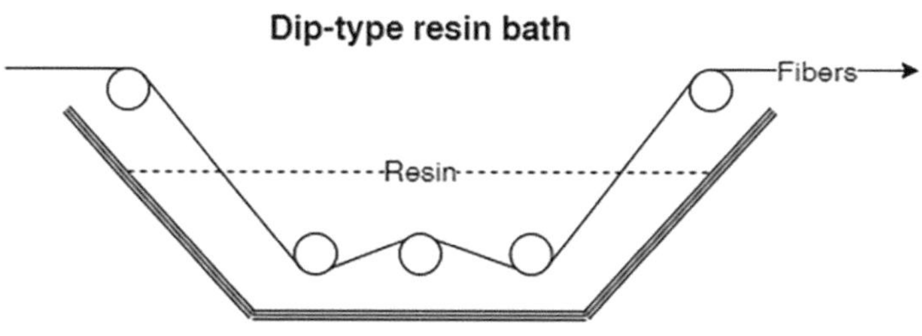

Fig. 1.19 Dip-type resin bath impregnation system

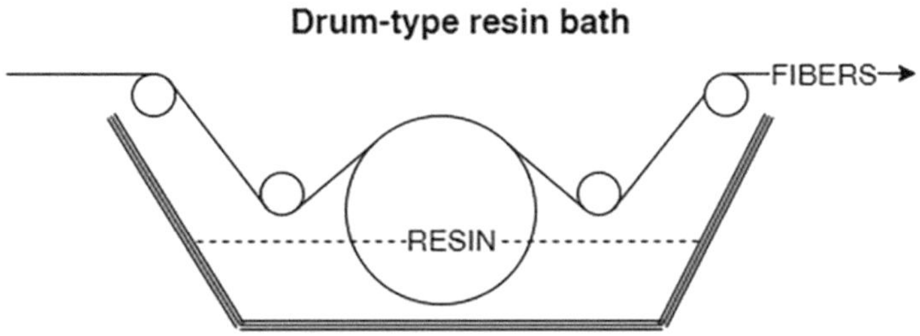

Fig. 1.20 Drum-type resin bath impregnation system

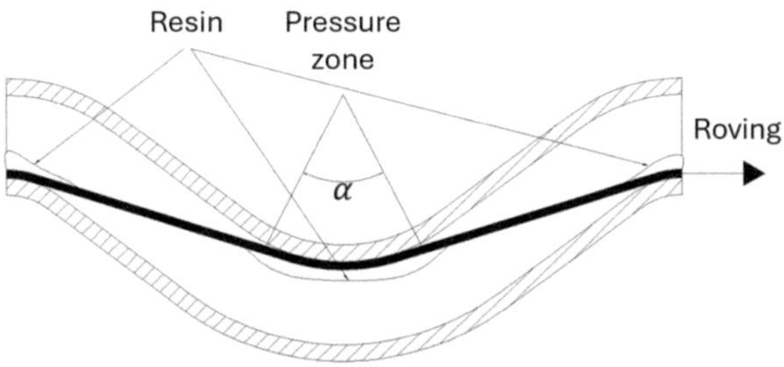

Fig. 1.21 Low pressure impregnation system

The systems, which used the pressure of the resin in the impregnation process, could be divided into low- and high-pressure techniques. Low-pressure systems use the geometry of the device to impregnate whole fibers through the cross section of the roving. Modification of the bending curve and radius used provide control over the pressure zone effect on the roving (Fig. 1.21).

The last type is the high-pressure impregnation system, which uses direct injection of the resin into the mold. The shape of the section in the die must be fitted to the applied matrix system. The resin used in the high-pressure application could be of higher viscosity compared to previous techniques. The other advantages are the reduction of resin waste, which increases work safety by reducing volatile substances in the air. The two types of high-pressure impregnation system are presented in Figs. 1.22 and 1.23.

1.2.5 Splicer

To ensure continuous manufacturing and reduce the time spent replacing bobbins with fibers, the splicers are commonly used in the composite industry, mainly in the pultrusion and filament winding process. The tool provides two separate rovings to add additional yarn of the fibers or replace the ending ones. The solution could be used for various types and Tex of reinforcement. To ensure the best quality of the joint, the air pressure should be adjusted for different materials according to

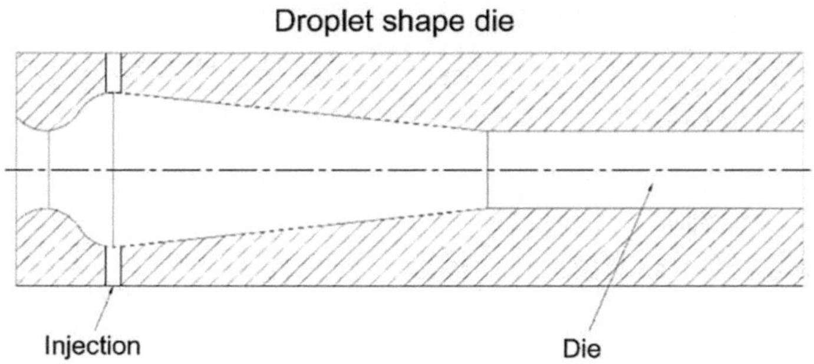

Fig. 1.22 High pressure impregnation system-droplet-shaped die

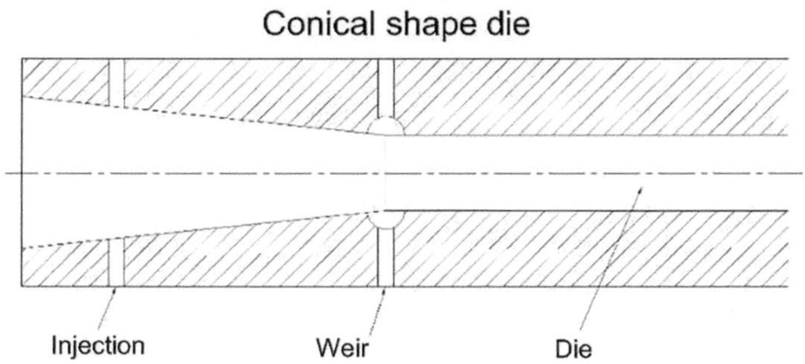

Fig. 1.23 High pressure impregnation system conical shape die

the manufacturer's recommendations. The connection of the two separate rovings is described in Fig. 1.24.

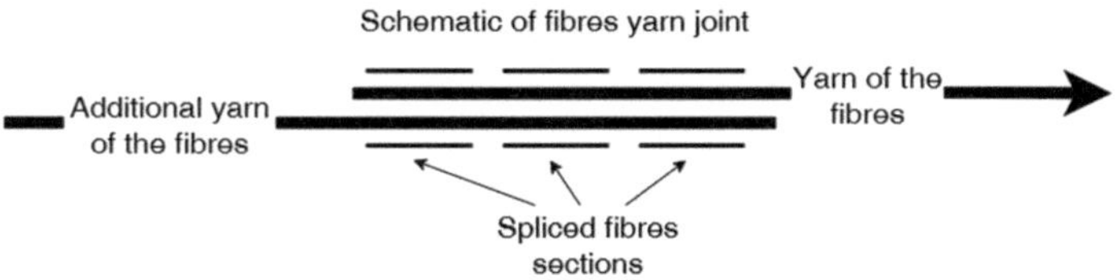

Fig. 1.24 Splicer tool to join two separate yarns of fibers

1.3 Application of FRP Rebars

Composite rebars are especially designed for demanding fields of application in harsh environments, where the structure must withstand permanently high mechanical and corrosive loads. The application of FRP bars could significantly reduce the maintenance cost and service life of the construction compared to conventional steel bars [31, 32]. Australia weather conditions present large amounts of aggressive agents such as seawater, relative humidity, and temperature. The market estimate said that construction companies used 2.5 million meters of FRP bar across the continent between 2012 and 2018. Structures using FRP profiles are located mostly near or inside the ocean, such as the Pinkenba Wharf or the Anthon's Landing Wynd-ham community jetty. GFRP bars are used to achieve maintenance-free infrastructure [33].

Civil Infrastructure

The largest commercial project in the world that uses GFRP rebars is the Jizan flood mitigation channel finished in 2022 with 23 km of length and 80 m of width. The site of construction is in the Red Sea, where floods are the result of seasonal rainfall and deficient drainage systems. The structure was created to reduce the damage caused by the water. The entire construction used 4800 km of rebars, of which one quarter was delivered with a bending state. The construction life is designed for 100 years and the annual repair cost is reduced by 10%.

The new industrial city project, financed by Saudi Arabia, King Salman Energy Park, announced the use of 700 thousand rebars in various delivery shapes, for buildings and sidewalks. One of the applications will be the Al Qurayyah bridge deck, where replacement of the steel reinforcement by GFRP bars provides a 74% mass reduction.

The high bending flexibility of the FRP rebars is used in various ambitious projects. The construction with complicated shapes, large radius corners such as skatepark in Alfred Cox Park, Gisborne demands application of the material with flex ability. GFRP bars provide a cost-effective correspondence with bending of the steel bars and labor work.

Bridges

The reduction in maintenance cost forced a growing number of agencies to select composite barbeds as the most effective long-term solution for bridge construction. The Floodway Bridge over the Red River in Winnipeg (Canada) is reinforced by 140 tons of GFRP rebar, making the structure the largest non-metallic reinforced concrete bridge in the world. The first arch bridge constructed with composite reinforcement is the Sumani Trace Bridge in Trinidad. The contractors used a building panel system, which allows them to install whole panels to fabricate the ceiling in one day. The concrete bridge constructed in Quebec (Canada) used composite bars as deck reinforcement. Applied instrumentation such as fiber optic sensors provides long-term measuring of loading condition to evaluate the effect of various factors such as impact loading, temperature, vibration. The conclusion obtained after a period of time shows that the performance of sand-coated GFRP bars is similar to that of their steel counterparts [34].

Seismic Application

The researcher conducted several studies to validate the FRP rebars under seismic conditions. The outline of the experiments presents that the concrete structures reinforced with composite bars exhibit a smaller residual displacement under low cyclic loading. The total cumulative dissipated energy is like the beam reinforced by traditional bar. Structures reinforced with GFRP bar materials have higher displacement, lower ductility, and stiffness compared to their steel counterparts in the same force loading condition. The properties cited could lead to a reduction in maintenance costs, which should be considered during cost evaluation [35–39].

Mining and Tunneling

Today, tunnels are mechanically excavated using tunneling machines (TBM). The pre-cast concrete segments are then installed to support the machine and resist temporary and permanent loads. The GFRP reinforcement is forced to be applied in tunnel construction to extend the serviceability of the structure, which is designed for more than 100 years of exploration. The advantages of replacing steel reinforcement are increasing durability, reducing concrete protective layers mainly in locations with a thin thickness, and electrical insulation of GFRP reinforcement. Segments with GFRP reinforcement could be easily demolished using the TBM machine due to the realization of lateral cross-passages, joining the new tunnel line. The realization of fully reinforced segments with GFRP bars is Milan Metro Line 4 (Italy) and IDRIS Doha Project (Qatar) [40, 41].

The composite rebars are also used to stabilize the working faces and anchor the roof in the mining. One of the non-conventional pros of the FRP bars is easy removal of the destroyed rebar from the debris, for example, after an explosion. Innovative innovations applied in the mining industry are GFRP pipe anchor bolts with injection systems to directly implement cement or two-component chemical anchor [42].

Marine Structures

The saline and alkaline environmental conditions provide a reduction of mechanical properties during long-term exposure. The FRP bars with dedicated polymer matrix and reinforcement with additional sand coating is one of the solutions used. Application of composite rebars provides enhanced flexural behavior, reduced crack spacing [43].

Acknowledgements The research was financed by the National Center for Research and Development of Poland (NCBiR) grant number LIDER/40/0219/L201010/18/NCBR/2019.

References

1. Mugahed Amran YH, Alyousef R, Rashid RSM, Alabduljabbar H, Hung CC (2018) Properties and applications of FRP in strengthening RC structures: a review. Structures 16:208–238
2. Feng G, Zhu D, Guo S, Rahman MZ, Jin Z, Shi C (2022) A review on mechanical properties and deterioration mechanisms of FRP bars under severe environmental and loading conditions. Cem Concr Compos 134:104758
3. Arrabiyeh PA, May D, Eckrich M, Dlugaj AM (2021) An overview on current manufacturing technologies: processing continuous rovings impregnated with thermoset resin. Polym Compos 42(11):5630–5655. Wiley
4. Nepomuceno E, Sena-Cruz J, Correia L, D'Antino T (2021) Review on the bond behavior and durability of FRP bars to concrete. Constr Build Mater 287:123042
5. Portnov G, Bakis CE, Lackey E, Kulakov V (2013) FRP reinforcing bars - designs and methods of manufacture (review of patents). Mech Compos Mater 49(4):381–400. Accessed 29 March 2024. https://doi.org/10.1007/s11029-013-9355-1
6. Cousin P, Hassan M, Vijay PV, Robert M, Benmokrane B (2019) Chemical resistance of carbon, basalt, and glass fibers used in FRP reinforcing bars. J Compos Mater 53(26–27):3651–3670. Accessed 15 Aug 2023. https://journals.sagepub.com/doi/full/10.1177/0021998319844306?casa_token=gAJG1twLRMcAAAAA%3A2TSkrPKWmjRayr_77-wz6hSAnSHqu4I80Os445BVkLJTO6oAxFplSX7kUApNFcG_S2m1oylUjCaf
7. Al-Mahmoud F (2018) 3.22 CFRP reinforcement rods. In: Comprehensive composite materials II, pp 578–591
8. Xiong Z, Zeng Y, Li LG, Kwan AKH, He SH (2021) Experimental study on the effects of glass fibres and expansive agent on the bond behaviour of glass/basalt FRP bars in seawater sea-sand concrete. Constr Build Mater 274:122100
9. Sonnenschein R, Gajdosova K, Holly I (2016) FRP composites and their using in the construction of bridges. Procedia Eng 161:477–482
10. El-Tahan M, Galal K, Hoa VS (2013) New thermoplastic CFRP bendable rebars for reinforcing structural concrete elements. Compos B Eng 45(1):1207–1215
11. Van De Velde K, Kiekens P (2001) Thermoplastic pultrusion of natural fibre reinforced composites. Compos Struct 54(2–3):355–360
12. Minchenkov K, Vedernikov A, Safonov A, Akhatov I (2021) Thermoplastic pultrusion: a review. Polymers 13(2):180. Accessed 08 Aug 2023. https://www.mdpi.com/2073-4360/13/2/180/htm
13. Zielonka P et al (2022) The effect of flame retardant—aluminum trihydroxide on mixed mode I/II fracture toughness of epoxy resin. Polymers (Basel) 14(20)
14. Shi XH, Li XL, Li YM, Li Z, Wang DY (2022) Flame-retardant strategy and mechanism of fiber reinforced polymeric composite: a review. Compos B Eng 233:109663
15. Romanovskaia N, Minchenkov K, Gusev S, Klimova-Korsmik O, Safonov A (2023) Effects of additives on the mechanical and fire resistance properties of pultruded composites. Polymers 15(17):3581. Accessed 28 March 2024. https://www.mdpi.com/2073-4360/15/17/3581/htm
16. Boukhili R, Boukehili H, Ben Daly H, Gasmi A (2006) Physical and mechanical properties of pultruded composites containing fillers and low profile additives. Polym Compos 27(1):71–81. Accessed 29 March 29. https://onlinelibrary.wiley.com/doi/full/https://doi.org/10.1002/pc.20127
17. Lobanov MV, Gulyaev AI, Babin AN (2016) Improvement of the impact and crack resistance of epoxy thermosets and thermoset-based composites with the use of thermoplastics as modifiers. Polym Sci Ser B 58(1):1–12. Accessed 29 March 2024. https://doi.org/10.1134/S1560090416010048
18. Perez RM et al (2007) Novel phosphorus-modified polysulfone as a combined flame retardant and toughness modifier for epoxy resins. Polymer (Guildf) 48(3):778–790
19. Karbhari VM (1998) Effect of internal mold release agent on the cure and property variation in resin transfer molding composites. J Mater Sci Lett 17(24):2061–2062. Accessed 29 March 2024. https://doi.org/10.1023/A:1006663612780

20. Duda S, Lesiuk G, Zielonka P, Stabla P, Lubecki M, Ziółkowski G (2021) Materials flexural pseudo-ductility effect in hybrid GFRP/CFRP bars under static loading conditions. https://doi.org/10.3390/ma14195608
21. Liu Y et al (2021) Experimental study on mechanical properties of novel FRP bars with hoop winding layer. Adv Mater Sci Eng 2021
22. Cui Y, Cheung MMS, Noruziaan B, Lee S, Tao J (2008) Development of ductile composite reinforcement bars for concrete structures. Mater Struct/Mater Constr 41(9):1509–1518
23. Cheung MMS, Tsang TKC (2010) Behaviour of concrete beams reinforced with hybrid FRP composite rebar
24. Tomasz MM, Szymon D, Bartosz B, Joanna W, Paweł Z, Grzegorz L (2022) Flexural and compressive residual strength of composite bars subjected to harsh environments. Eng Fail Anal 133:105958. Accessed 20 Aug 2023. https://doi.org/10.1016/j.engfailanal.2021.105958
25. Galos J. Microwave processing of carbon fibre polymer composites: a review
26. Zhilyaev I et al (2020) Controlled curing of thermoset composite components using infrared radiation and mathematical modelling. Accessed 30 March 2024. https://doi.org/10.1016/j.compstruct.2020.113224
27. Alpay YO, Uygur I, Kilincel M (2020) On the optimum process parameters of infrared curing of carbon fiber-reinforced plastics. Polym Polym Compos 28(6):433–439. Accessed 30 March 2024. https://journals.sagepub.com/doi/full/https://doi.org/10.1177/0967391119900711
28. Chaparala SAC et al (2023) Characterisation of curing of Vinyl Ester resin in an industrial pultrusion process: influence of die temperature. Polymers 15(18):3808. Accessed 30 March 2024. https://www.mdpi.com/2073-4360/15/18/3808/htm
29. Gadam SUK, Roux JA, McCarty TA, Vaughan JG (2000) The impact of pultrusion processing parameters on resin pressure rise inside a tapered cylindrical die for glass-fibre/epoxy composites. Compos Sci Technol 60(6):945–958
30. Krasnovskii A, Kazakov I, Krasnovskii A, Kazakov I (2012) Determination of the optimal speed of pultrusion for large-sized composite rods. J Encapsul Adsorp Sci 2(3):21–26. Accessed 30 March 2024. http://www.scirp.org/Html/1-1060023_23073.htm
31. Benmokrane B, Mohamed H, Ahmed E (2024) Recent developments of FRP bars as internal reinforcement in concrete structures & field applications. Accessed 02 April 2024. http://www.claisse.info/Proceedings.htm
32. Mohamed HM (2020) Innovative field applications of fiber reinforced polymer composite reinforcing bars in civil engineering infrastructures. Appl Sci Uni J 4(2):49–55. Accessed 02 April 2024. https://doi.org/10.18576/jasu/040202
33. Manalo AC, Mendis P, Bai Y, Jachmann B, Sorbello CD (2021) Fiber-reinforced polymer bars for concrete structures: state-of-the-practice in Australia. J Compos Constr 25(1). Accessed 30 March 2024. https://orcid.org/0000-0003-0493-433X
34. Rahman B, Saadatmanesh H, Rizkalla H (1998) Field application of FRP composite bars as reinforcement for bridge decks
35. Mady M, El-Ragaby A, El-Salakawy E (2011) Seismic behavior of beam-column joints reinforced with GFRP bars and stirrups. J Compos Constr 15(6):875–886. Accessed 02 April 2024. https://doi.org/10.1061/%28ASCE%29CC.1943-5614.0000220
36. Aliasghar-Mamaghani M, Khaloo A (2019) Seismic behavior of concrete moment frame reinforced with GFRP bars. Compos B Eng 163:324–338
37. Muntasir Billah AHM, Shahria Alam M (2012) Seismic performance of concrete columns reinforced with hybrid shape memory alloy (SMA) and fiber reinforced polymer (FRP) bars. Constr Build Mater 28(1):730–742
38. Su C, Wang X, Ding L, Chen Z, Liu S, Wu Z (2021) Experimental study on the seismic behavior of seawater sea sand concrete beams reinforced with steel-FRP composite bars. Eng Struct 248:113269
39. Sun Z-Y, Wu G, Wu Z-S, Zhang M (2011) Seismic behavior of concrete columns reinforced by steel-FRP composite bars. J Compos Constr 15(5):696–706. Accessed 02 April 2024. https://doi.org/10.1061/%28ASCE%29CC.1943-5614.0000199

40. Spagnuolo S, Meda A, Rinaldi Z, Nanni A (2017) Precast concrete tunnel segments with GFRP reinforcement. J Compos Constr 21(5):04017020. Accessed 30 March 2024. https://doi.org/10.1061/%28ASCE%29CC.1943-5614.0000803
41. Hosseini SM, Mousa S, Mohamed HM, Eslami A, Benmokrane B (2022) Experimental and analytical study on precast high-strength concrete tunnel lining segments reinforced with GFRP bars. J Compos Constr 26(5):04022062. Accessed 30 March 2024. https://doi.org/10.1061/%28ASCE%29CC.1943-5614.0001257
42. Borgmeier E (2014) Fibre-reinforced plastic (FRP) – an innovative approach to further development of rock bolt technology and the development of new outlets in tunnel and structural engineering. Min Rep 150(4):200–207. Accessed 30 March 2024. https://doi.org/10.1002/mire.201400027
43. Firdows M, Zakkaria M, Shanmugam P (2024) Flexural behavior of fiber-reinforced concrete beams with GFRP rebars under marine environmental conditions. Int J Adv Manuf Technol. Accessed 02 April 2024. https://doi.org/10.1007/s00170-023-12690-6

Chapter 2
Mechanical Properties of FRP Bars

Szymon Duda, Paweł Zielonka, Michał Smolnicki, Paulina Mayer-Trzaskowska, Tomasz Socha, Arkadiusz Denisiewicz, Krzysztof Kula, and Grzegorz Lesiuk

Abstract From a design point of view, the mechanical behavior of the material is crucial to choosing the appropriate material and geometry for the assumed boundary conditions. This chapter presents experiments investigating the mechanical performance of fiber-reinforced bars for concrete reinforcement, but is not limited to this aspect alone. Although composite materials have been well-known for decades, they are still not commonly used in civil structures. It is caused by limited standards for mechanical testing as well as different design strategies in comparison to commonly used engineering materials. This chapter includes the results of the authors' experimental campaign to investigate the mechanical properties of bars manufactured at Wroclaw University of Science and Technology by the modified pultrusion process.

Keywords Mechanical performance · FRP bars · Unidirectional composites · Elastic properties

2.1 Introduction

Engineering is nowadays changed by utilizing modern, other than metallic materials, which have an increasing presence every year. One of the groups covers continuous fiber materials, which find their place in almost every market branch including automotive, chemistry, aerospace, or civil engineering. Considering structural engineering, they replace many steel components, especially concrete reinforcement in offshore structures.

The structural performance of concrete structures with FRP bars is highly dependent on the mechanical performance of the reinforcement bars. Therefore, a comprehensive examination of the mechanical behavior of the components is crucial and

---

S. Duda (✉) · P. Zielonka · M. Smolnicki · P. Mayer-Trzaskowska · G. Lesiuk
Faculty of Mechanical Engineering, Wrocław University of Science and Technology, Wrocław, Poland
e-mail: Szymon.duda@pwr.edu.pl

T. Socha · A. Denisiewicz · K. Kula
Institute of Building, University of Zielona Góra, Zielona Góra, Poland

G. Lesiuk et al. (eds.), *Composite Rebars for Reinforced Concrete Structures*, Structural Integrity 34, https://doi.org/10.1007/978-3-031-96292-9_2

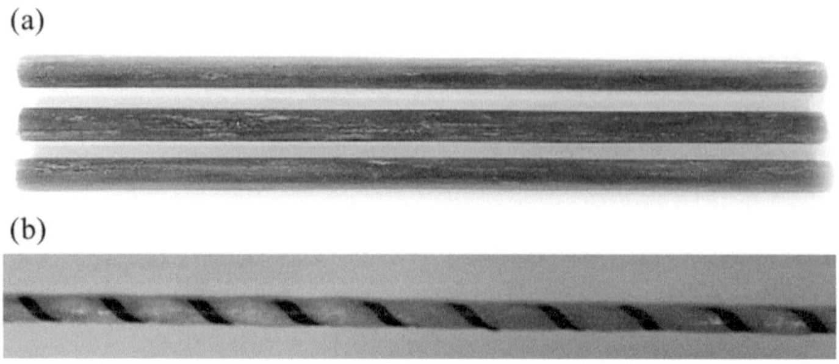

Fig. 2.1 Pultruded composite bars, **a** unidirectional rod, **b** hybrid composite rebar for concrete reinforcement manufactured using a modified pultrusion method

essential. Unlike metallic bars, composite bars require more sophisticated experimental procedures due to material anisotropy and their sensitivity to compression stresses that can be induced in the process of gripping.

This chapter discusses in detail the mechanical tests along with the results obtained for pultruded composite rods (Fig. 2.1a) and rebars for concrete reinforcement (Fig. 2.1b). Differences in manufactured bars are presented in Fig. 2.1 (see the additional fibers on the surface of the rods in the latter).

The design of members or structures requires an understanding of the mechanical behavior of the used materials. To describe the mechanical performance of those materials, the elastic properties need to be determined. When composite materials are considered, which are the combination of at least two constituents, the examination of elastic properties is more challenging and time-consuming. There are two approaches for defining these properties (which include Young's modulus E, Poisson ratio ν, and shear modulus G). First, the mechanical properties of subcomponents (fibers, matrix) need to be investigated; it allows for defining the effective elastic properties of fiber-reinforced composite material using the micromechanical approach such as Rule of Mixture [1], Abolin'sh [2], or others. However, it requires very sophisticated devices and testing machines, such as testing machines for single fibers, and it also requires the determination of fiber content. Considering the properties of a single fiber, the most challenging is the Poisson ratio, and there is no widely accepted method for defining it. This approach is presented in [3] for filament-wound members.

Another approach contains a macro-scale investigation that covers experimental tests on composite material systems. Macroscale experimental approaches are addressed, namely, using four-point bending tests, short beam tests, and tensile tests. Some national and international standards have already been issued. They provide knowledge of the experimental tests; however, they do not provide precise information about sample preparation if some deviation in geometry or material is considered. As a summary, some recommendations and tips will be written based on the experience of the authors on the complex mechanical analysis of the FRP bars.

2.2 Static Mechanical Tests

Application of composite bars in structural members requires different design approaches from those well-known for metallic materials due to their anisotropy and inhomogeneity. This process is highly complex to provide reliability and safety. During the last decade, considerable effort has been put into investigating the mechanical performance of composite bars for concrete reinforcement, crucial in some applications, such as concrete members affected by harsh environments.

Due to this fact, several standards that define the methodology for material characterization appeared, and some of them are presented in Table 2.1. This subchapter describes basic mechanical tests such as the tensile/compression or flexural test. It follows also the comments and suggestions of the standard authors.

2.2.1 *Static Tensile Test*

The tensile test is the most common mechanical test performed under static or cycling loading conditions. In terms of composite bars, there are several standards (ASTM

Table 2.1 Standards covering static experiments for composite bars

Test	Standards
Tensile	ASTM D7205 [4], ISO 10406-1 [5], EN 17129 [6]
Tensile creep	ASTM D7337 [7]
Transverse shear	ASTM D7617 [8]
Apparent horizontal shear strength	ASTM D4475 [9]
Flexural strength	ASTM D4476 [10]

Table 2.2 Sample configuration including information about dimension and surface preparation

Sample	Surface preparation	Diameter [mm]	L [mm]	L_a [mm]
BFRP1	Primer and quartz sand (1–3 μm)	6	300	400
GFRP2	Primer	6	300	400
GFRP3	Primer and quartz sand (3–6 μm)	6	300	400
GFRP4	Primer and quartz sand (1–3 μm)	6	300	400
BFRP5	Sandblasting	6	300	400
BFRP6	Primer and quartz sand (3–6 μm)	6	300	400
BFRP7	Primer	6	300	400
GFRP8	Sandblasting	6	300	400
GFRP8	Primer and quartz sand (1–3 μm)	10	300	500

D7205, CAN/CSA S806, CNR-DT 206, GOST 31938, ISO 10406-1, EN 17129, JSCE-E 531) that provide a procedure for conducting the tensile test. The focus here is on the ASTM D7205 standard.

The tensile test for composite bars is quite problematic due to the sample size. The mentioned standard gives the recommended lengths of the specimens and steel tubes with respect to diameter. Due to the material sensitivity to compression and to avoid initiation and fracture in a gripping part, a special anchor needs to be provided. A thick-walled steel tube is an alternative for an anchor, which is on both sides of the sample for a specific length to provide sufficient bond strength between the filler—bar and the filler–steel tube. As a filler material, the standard suggests epoxy resin or a mix of quartz sand and epoxy resin. The emphasis needs to be placed on the bar alignment concerning the applied tensile force; it can be done in various ways; however, plastic caps with central holes provide good results.

It is very important to keep the recommended length and dimensions of the anchors; otherwise, the pull-out can be observed. Moreover, bar surface preparation is very important, since the adhesion between the bar and the matrix is weaker than between the steel and the matrix. The authors' experience suggests sandblasting both gripped parts provides a rough bar surface which bonds better to the filler.

The sample size requires a testing machine with a large space between the lower and upper grips. Moreover, due to the relatively heavy size of the anchor (steel tube diameter from 35 up to 75 mm), the grips need to also be capable of holding and carrying the load. The experimental setup is shown in Fig. 2.2. For strain measurement, an optical method such as digital image correlation can be applied. It can reach an area larger than the extensometer and provides information about the local stress concentration.

The authors conducted an experimental campaign to determine the tensile properties, including the axial stiffness of the manufactured rods and rebars. Data on the configuration of the sample are presented in Table 2.1, where BFRP—basalt fiber reinforced polymer, GFRP—glass fiber reinforced polymer, L—gauge length, L_a—anchor length.

As can be seen, if the 10 mm diameter is considered, the anchor length is 500 mm, which in total gives the sample length 1300 mm. As mentioned, it requires a larger testing machine, in the case of the authors' experiment it was conducted on INSTRON 8804 $\pm$ 500 kN and Dantec Q-400 system was used for strain monitoring using digital image correlation. Moreover, in Fig. 2.2 the blue steel plate can be seen; it is a thick metal plate, used to grip one side of the specimen to increase the space, instead of using a large hydraulic grip in the machine.

2.2.2 *Static Compression Test*

Composite materials exhibit brittle behavior, which generates some challenges, especially during compressive load. First, the compressive load cannot be carried by

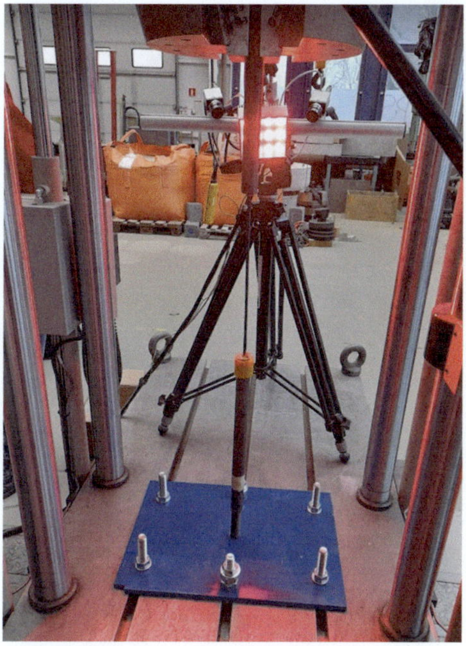

Fig. 2.2 Tensile test of the FRP bar according to ASTM D7205. Experiment equipped with the digital image correlation

the fiber itself, so the load is carried mainly by the matrix, especially in unidirectional fiber composites. Second, the technology causes material discontinuity, which impacts the compressive strength significantly. All these aspects make those material systems sensitive to compressive load, and it is still an issue, especially under cycling load.

There is no standard for compression loading conditions, this loading case beside the issues mentioned before can cause buckling on a macro- or micro-scale. An alternative to that approach is radial compression, which was carried out by us for two different materials, e.g., carbon (CFRP) and glass (GFRP) reinforced polymers, including various lengths. Considering the radial compression, there is no possibility of buckling, and only the matrix will carry the applied load. The lengths of three different specimens were chosen to investigate their effect on the final results, the dimensions of the specimens are listed in Tables 2.3 and 2.4 and the specimens are shown in Fig. 2.3. For the static radial compression test, a servo-hydraulic testing machine with 25 kN was applied, the compression plates with the specimen in between are given in Fig. 2.4.

Table 2.3 Average dimensions of glass fiber reinforced polymer rods

Length [mm]	Diameter [mm]
10.01	10.01
30.03	10.01
49.96	10.01

Table 2.4 Average dimensions of carbon fiber reinforced polymer rods

Length [mm]	Diameter [mm]
10.47	9.99
29.90	10.00
49.78	9.99

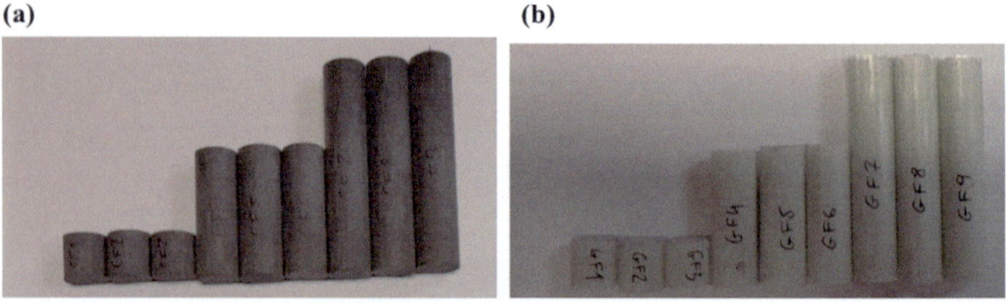

Fig. 2.3 Composite bars used in compression experiments **a** CFRP, **b** GFRP

Fig. 2.4 Specimen placed between compression plates

2.2.3 Bending Tests

Considering the reinforcing bars the flexural behavior is a major aspect to consider while designing. Bending is one of the major loading cases, especially in civil structures, in that there is a great need to investigate this behavior. In experimental mechanics, mainly 3-point or 4-point bending tests are considered. In terms of composite unidirectional bars, there is an ASTM D4476 standard, which covers the aspect of bending. It gives the formula for maximum stress and strain during bending, as well as allows for the calculation of the tangent modulus of elasticity. As discussed before, composite materials are highly sensitive to compressive stress, (which occurs as a part of bending). Thus it is recommended to reduce the area under

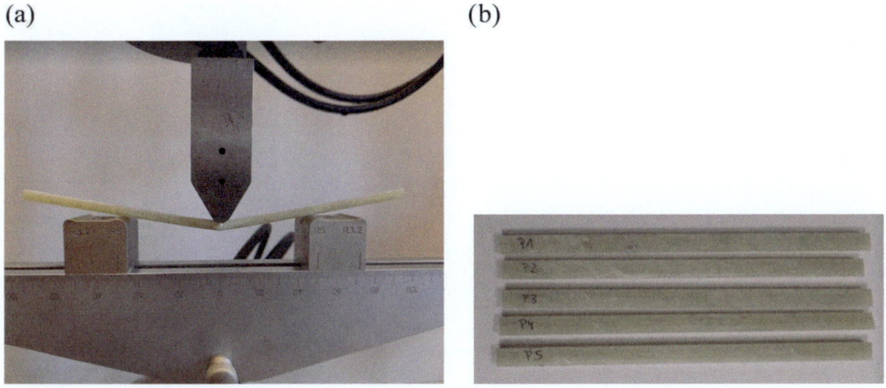

Fig. 2.5 **a** Experimental set-up for the three-point bending test with the half-circular specimens, **b** exemplary set of 5 half-circular specimens (series P) used in the bending test

compressive stress by cutting the composite bar along the bar's axis in a way that the resultant cross-section is less than half of the original circular cross-section. The idea of this method is described in detail in [11].

The authors performed the test for GFRP bars with a diameter of 10 mm in the testing system with a 25 kN load cell and displacement control mode with a rate of 3 mm/min. The experimental setup and prepared specimens are shown in Fig. 2.5. It is worth mentioning that the most challenging thing in this experiment is to prepare specimens that are cut along the axis; it is becoming more difficult when smaller diameters are considered. The bars were cut with a diamond wire saw.

2.2.4 *Shear Tests*

Along with bending, shear strength should also be considered, especially in anisotropic materials. The effect of shear stress on the members is investigated during the short beam test. This experiment is well described in ASTM D4475, and the experimental setup is given in Fig. 2.6. As can be seen, the fixture for the three-point bending test should be used, however, the specimen length is much smaller to increase the impact of shear stresses.

The short-beam test was conducted for the following sample batches: GFRP 10, 6, BFRP 6, Hybrid 8, where the hybrid bar is a combination of the core of basalt fibers and the glass fibers surrounding it. The span for the experiment is defined by multiplying the diameter by 3, so the specimens should be slightly longer than this. The test was controlled by a constant displacement rate of 1.3 mm/min. For this test as well as for bending, special supports need to be provided, which are highlighted in the standards, it is important to properly place and reduce the friction at the supports during flexural tests. From the results, the apparent shear strength can be calculated,

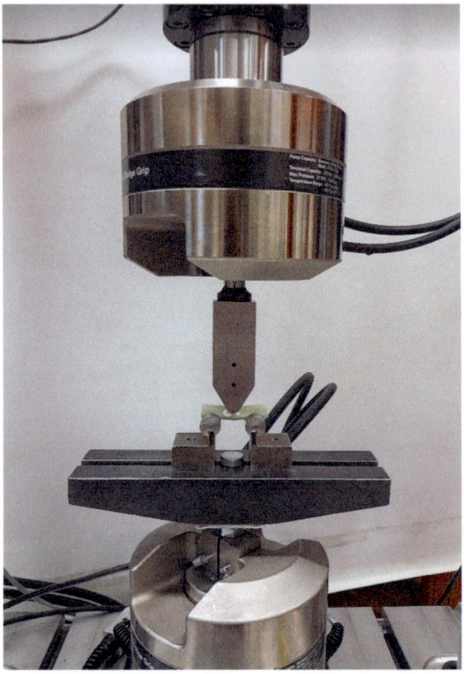

Fig. 2.6 Short beam test for shear determination—experimental setup

which may be directly correlated with the mechanical performance of the matrix and the adhesion strength between the fibers and the matrix. These aspects are significant if cycling loading is concerned. The apparent shear strength is calculated using the following formula:

$$S = 0.849P/d^2 \tag{2.1}$$

where S—apparent shear strength [MPa], P—breaking load [N], and d—specimen diameter [mm].

2.3 The Experimental Results of the Authors

This subchapter covers the authors' experimental data obtained during the aforementioned mechanical experiments of pultruded composite bars at the Wroclaw University of Science and Technology.

2.3.1 Tensile Behavior

The tensile behavior of structural members is crucial in terms of design, and considering the metallic materials, it is also one of the easiest experiments to conduct. As mentioned in terms of composite bars, it is challenging. In Fig. 2.7, the results of the tensile tests are shown for different bars with an analysis of Young's modulus. The results clearly show that the smooth composite bars in both cases reinforced with basalt and glass fibers exhibit brittle behavior.

The results obtained are presented in Table 2.5. In some cases, the tensile strength was higher than the adhesion strength between bar and resin and resin and steel tube. If this happened, the pull-out of the bar from the anchor was observed and was excluded the specimen from the analysis. Specimens that presented this type of behavior are denoted with the "pull-out" in the table.

The delivered Young's modulus for BFRP is 48.3 GPa with standard deviation 0.83 GPa, and for GFRP is 44.1 GPa with standard deviation 0.57 GPa; as can be seen, the stiffness of these composite bars is much lower than for metallic bars; however, the UTS is very high reaching even 1400 MPa. If the UTS is divided by the density of the composite materials, and compared with metallic materials, the ratio is higher, which makes those materials very attractive in many applications, especially lightweight.

2.3.2 Results—Modulus of Elasticity in Bending

If Young's modulus is considered, only a tensile test should be performed to deliver this value. However, the case of bending loading is a combination of tension and compression, in that the modulus of elasticity in bending can be calculated but it cannot be directly associated with Young's modulus. The calculations of this modulus are given by ASTM D4476, and they are given below:

$$E_b = \frac{PL^3}{38YI} \tag{2.2}$$

where E_b—modulus of elasticity in bending [MPa], P—load at a given point on the load–displacement plot [N], L—support span [mm], I—moment of inertia [mm^4], Y—maximum deflection [mm].

Table 2.6 shows the results obtained, the mean value of modulus is equal to 79.9 GPa with standard deviation of 0.81 GPa. This value is valid for the sample made from glass fibers with a diameter of 10 mm, however, later cut through the bar axis. It is important to mention that the properties of the material in terms of composite materials depend on the size and fiber content. If the size is concerning, the scale effect needs to be considered; in terms of those materials, this effect is more pronounced, the larger the larger volume of defects. The significant impact on

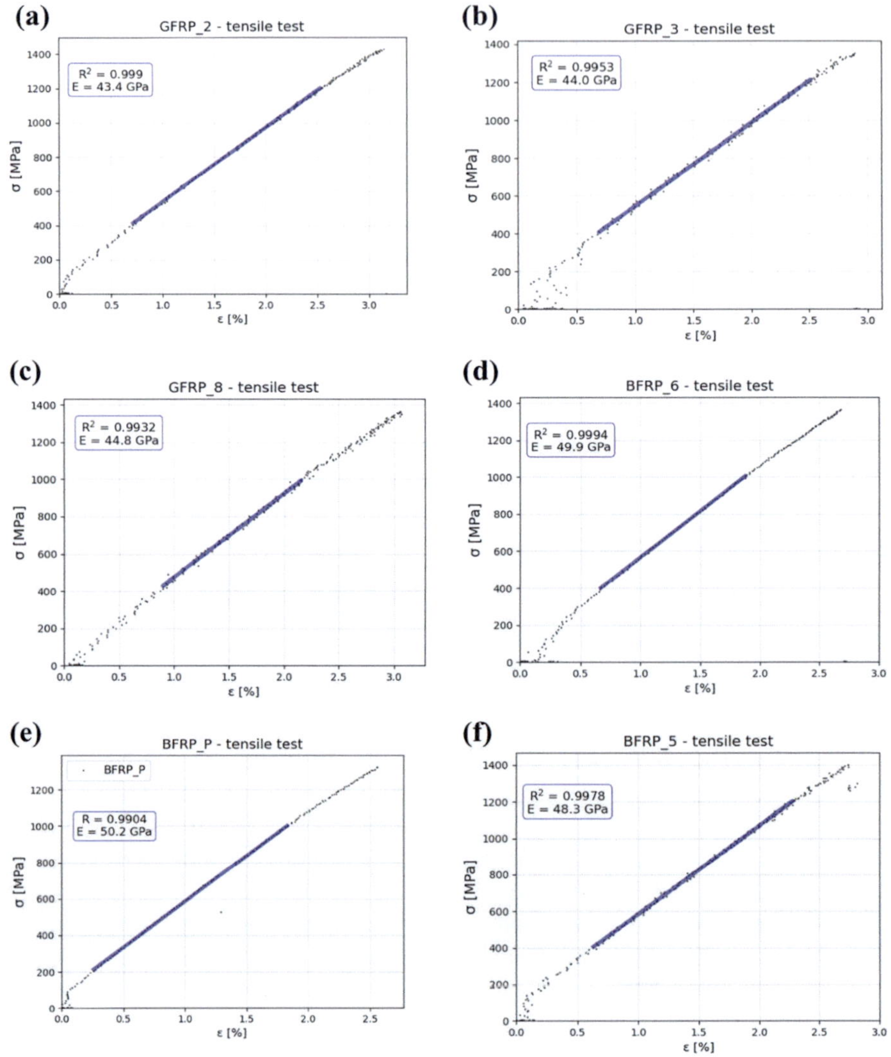

Fig. 2.7 Stress–strain results for **a**–**c** rods with glass fiber reinforcement and **d**–**f** rods with basalt fiber reinforcement. For each stress–strain characteristic Young Modulus was determined using linear regression

strength properties and elastic properties has fiber content, which means if the fiber content is not equal, the laminates or composite members will not provide the same mechanical performance.

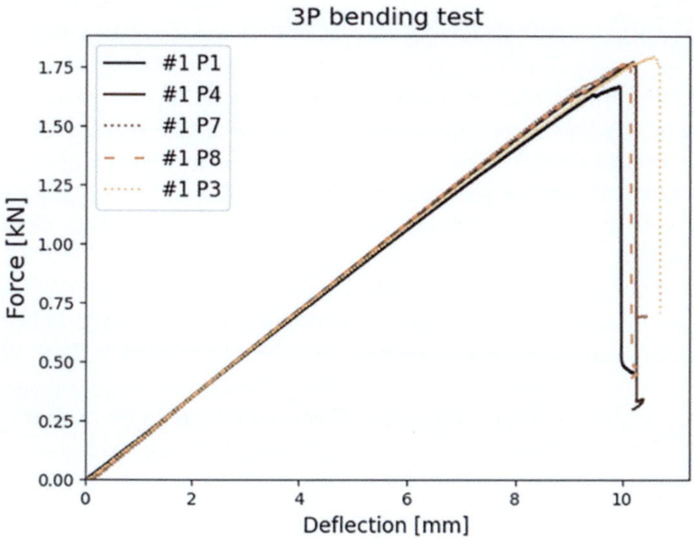

Fig. 2.8 Flexural behavior for GFRP bars

Table 2.5 Young's modulus data for the composite bars investigated

Sample no	Surface preparation	Diameter	E [GPa]
BFRP1	*Primer and quartz sand*(1–3 μm)	*6*	*Pull-out*
GFRP2	Primer	6	43.4
GFRP3	Primer and quartz sand (3–6 μm)	6	44.0
GFRP4	*Primer and quartz sand*(1–3 μm)	*6*	*Pull-out*
BFRP5	Sandblasting	6	48.3
BFRP6	Primer and quartz sand (3–6 μm)	6	49.9
BFRP7	Primer	6	50.2
GFRP8	Sandblasting	6	44.8
GFRP9	*Primer and quartz sand*(1–3 μm)	*10*	*Pull-out*

Table 2.6 Flexural moduli obtained from the bending test

Sample no	Modulus of elasticity in bending [GPa]
P1	78.5
P2	80.2
P3	80.7
P4	80.5
P5	79.4

2.3.3 Results—Apparent Shear Strength

The nature of the material subjected to shear stress during the short beam test is shown in Fig. 2.9, in this test the material behavior is governed by the polymeric matrix and its adhesion to fibers, the maximum force increases with the diameter, however, the most unstable post-failure behavior is shown in case of C-bars so for GFRP bar with 10 mm diameter, which may suggest an impact of the defects on the performance.

The values for apparent shear strength are given in Table 2.7, and the largest value was obtained for the hybrid bar, which is a combination of basalt and glass fibers, if the shear stresses are considered, the highest value is expected near the neural axis of bending, in the case of H-bars, the core is reinforced with basalt fibers, which exhibit greater shear strength than glass fibers, which might be caused by various fiber preparation, moreover, basalt fibers are natural, which may also impact this value.

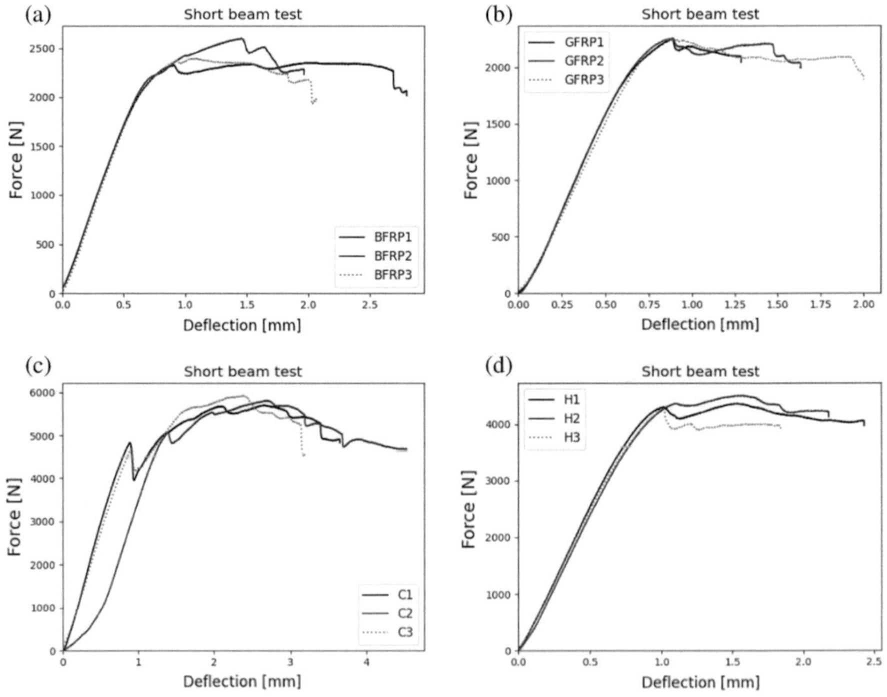

Fig. 2.9 Results of the short bending test for **a** BFRP, **b** GFRP, **c** C bars (GFRP Φ10), and **d** H bars (hybrid)

Table 2.7 Comparison of the results of apparent shear strength for various material systems

Sample no	Apparent shear strength [MPa]
H–bars	58.1
C–bars	49.3
GFRP	53.2
BFRP	57.8

2.3.4 Results—Radial Compression

The results for radial compression are presented in terms of 3 curves for every material length (CFRP and GFRP). At first glance, the initial stage of the plot might be considered, which reflects the alignment of a rod in compression plates. It can be seen that the maximum force is higher for CFRP bars, which suggests that there is better adhesion between matrix and carbon fibers or different fiber distribution and content since the carbon fibers possess a smaller diameter than the glass fibers; if the residual behavior is considered, there is randomness, which can be the results of fiber distribution in specimen, defects, or degree of polymerization since this test exhibits mainly the performance of matrix (Fig. 2.10).

2.4 Final Remarks

The comparison of the results obtained is given in Table 2.8, some of the parameters are not available due to the limitations of the testing systems or they are not available in the literature. As can be seen, Young's modulus for composite rods is in the range of 40–60 GPa for fiber content between 60 and 75%, if metallic bars are considered, there is a difference of more than 4 times; however, along with stiffness, the UTS value is very important and in this case, the range is higher than for typical steel bars. The composite bars exhibit even up to 1400 MPa UTS for the author composite bars. Considering the apparent shear strength, it is in the range of 40–60 MPa. The higher this strength, the greater the resin and adhesion performance of the composite bars.

This chapter shows the mechanical experiments devoted to the description of the mechanical performance of the composite bars in various ways. The most relevant test is the tensile test, which gives Young's modulus and UTS, however, if the application requires high-quality products, other tests can be in the area of interest. Composite bars despite the lower stiffness can be applicable in environments where metallic bars are affected by corrosion because they do not experience corrosion.

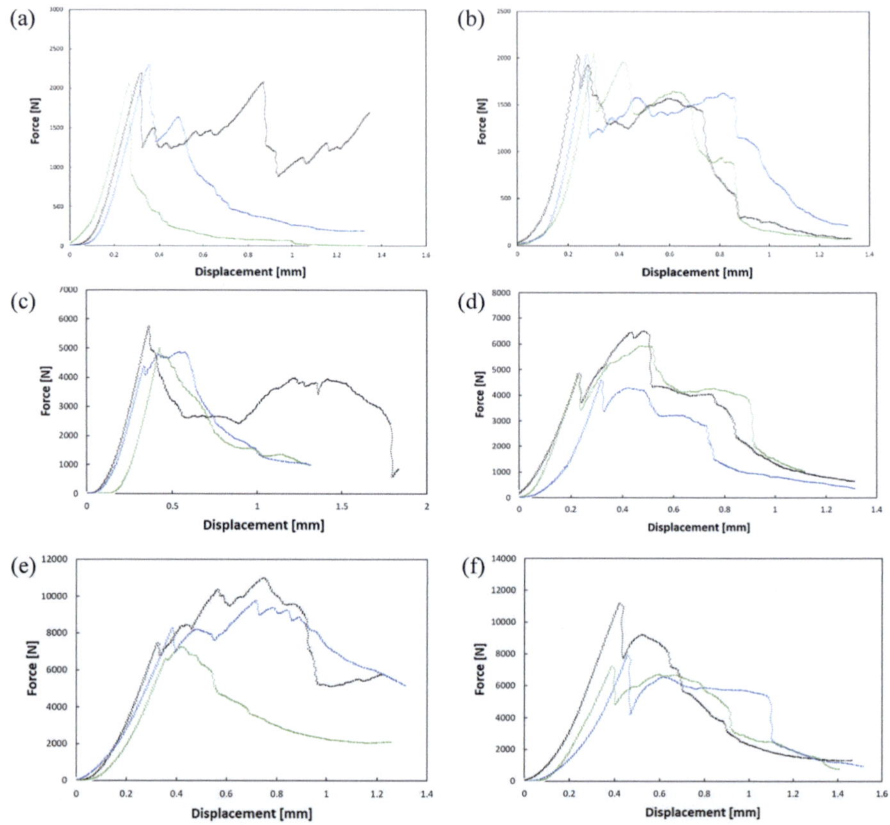

Fig. 2.10 Radial compression behavior for **a** CFRP L = 10 mm, **b** GFRP L = 10 mm, **c** CFRP L = 30 mm, **d** GFRP L = 30 mm, **e** CFRP L = 50 mm, **f** GFRP L = 50 mm

Table 2.8 Comparison of mechanical properties for various composite bars for data from the own and literature

	GFRP ϕ10	GFRP ϕ6	BFRP ϕ 6	HYBRID ϕ 8	Arczewska et al. [11]—GFRP	Wiater et al. [12]—GFRP	Lu et al. [13]—GFRP
Young's modulus [GPa]	N/A	44	50	N/A	**60**	**40**	41
Tensile Strength [MPa]	N/A	**1380**	1360	N/A	1000	971	1164
Apparent shear strength [MPa]	49	53	57	**58**	N/A	42	N/A
Modulus of Elasticity in Bending [GPa]	**80**	N/A	N/A	N/A	N/A	N/A	N/A
Fiber content (%)	60	62	62	N/A	75	69	75

Acknowledgements The research was financed by the National Center for Research and Development of Poland (NCBiR) grant number LIDER/40/0219/L-10/18/NCBR/2019.

References

1. Gooch JW (2011) Law of mixtures. In: Encyclopedic dictionary of polymers, pp 421–421. https://doi.org/10.1007/978-1-4419-6247-8_6810
2. Abolin'sh DS (1965) Compliance tensor for an elastic material reinforced in one direction. Polym Mech 1(4):28–32. https://doi.org/10.1007/BF00858888/METRICS
3. Duda S et al (2024) Experimental characterization and modeling of cylindrical CFRP structures under quasi-static multiaxial loading conditions. Thin-Walled Struct 195:111364. https://doi.org/10.1016/j.tws.2023.111364
4. ASTM D7205/D7205M-21 (2021) Standard test method for tensile properties of fiber reinforced polymer matrix composite bars, vol i, p 13. https://doi.org/10.1520/D7205
5. ISO 10406-1:2015—Fibre-reinforced polymer (FRP) reinforcement of concrete—Test methods—Part 1: FRP bars and grids. Accessed 9 Dec 2024. https://www.iso.org/standard/63657.html
6. EN 17129:2018—Continuous-fibre-reinforced plastic composites—Pultruded unidirectional rods. Accessed 9 Dec 2024. https://standards.iteh.ai/catalog/standards/cen/d433d2ab-5db4-40df-a9a7-ed87f206abd9/en-17129-2018?srsltid=AfmBOorO9aYAkBMp-jwVe73hdbdjd2BnydjCrv3QhOzu1eFMGfkJBRxH

7. Test method for tensile creep rupture of fiber reinforced polymer matrix composite bars. ASTM International, West Conshohocken, PA. https://doi.org/10.1520/D7337_D7337M-12R19
8. Test method for transverse shear strength of fiber-reinforced polymer matrix composite bars, 01 Aug 2017, ASTM International, West Conshohocken, PA. https://doi.org/10.1520/D7617_D7617M-11R17.
9. Test method for apparent horizontal shear strength of pultruded reinforced plastic rods by the short-beam method, 01 Jul 2021. ASTM International, West Conshohocken, PA. https://doi.org/10.1520/D4475-21
10. Standard Test Method (2009) Standard test method for flexural properties of fiber reinforced pultruded plastic. Practice i(December 2003):1–5. https://doi.org/10.1520/D4476
11. Arczewska P, Polak MA, Penlidis A (2019) Relation between tensile strength and modulus of rupture for GFRP reinforcing bars. J Mater Civ Eng 31(2):04018362. https://doi.org/10.1061/(ASCE)MT.1943-5533.0002575
12. Wiater A, Siwowski T (2020) Comparison of tensile properties of glass fibre reinforced polymer rebars by testing according to various standards. Materials 13(18):4110. https://doi.org/10.3390/MA13184110
13. Lu C, Yang Y, He L (2021) Mechanical and durability properties of GFRP bars exposed to aggressive solution environments. Sci Eng Compos Mater 28(1):11–23. https://doi.org/10.1515/SECM-2021-0002/MACHINEREADABLECITATION/RIS

Chapter 3
Damage Prediction Models

Michał Smolnicki and Szymon Duda

Abstract FRP rebars present linear elastic behavior until failure. Their lack of plasticity can lead to sudden failure or significant loss of stiffness. Thus, establishing reliable predictions of its behavior is of utmost importance. The first approach to establish a failure criterion for FRP was based on an extension of isotropic criteria (von Mises for example) to include the orthotropic nature of these materials. The most general version of such criteria is the Tsai-Wu criterion. These types of criteria can predict FRP failure, however without taking into account different failure modes. Within the second generation of failure criteria, which are focused on combining the influence of different failure modes the most well-known are maximum stress/strain and Hashin-Rotem criteria. Another aspect of predicting damage in FRP materials is post-failure behavior. Mentioned criteria are used only to determine the end of the elastic behavior and after that, there is the need to characterize the mechanical response of the material to further loading—so the progression of the damage. Generally, some evolution laws need to be defined—starting from reducing material properties to lower, predefined values and coming to more complex approaches such as defining such laws based on fracture energy dissipation. This chapter also addresses the application of criteria to predict the failure of FRP rebars. Different potential approaches to defining post-failure behavior in such materials will be also analyzed. Finally, the potential usage of mentioned models in numerical simulations will be discussed.

Keywords Damage initiation · Damage propagation · Composites · Damage prediction · FEA

Abbreviations

RVE Representative volume element

M. Smolnicki (✉) · S. Duda
Faculty of Mechanical Engineering, Wroclaw University of Science and Technology, Wroclaw, Poland
e-mail: michal.smolnicki@pwr.edu.pl

G. Lesiuk et al. (eds.), *Composite Rebars for Reinforced Concrete Structures*, Structural Integrity 34, https://doi.org/10.1007/978-3-031-96292-9_3

X	Allowable strength in the longitudinal direction in tension
Y	Allowable strength in the transversal direction in tension
S, Q, R	Allowable in-plane shear strength
X'	Allowable strength in the longitudinal direction in compression
Y'	Allowable strength in the transversal direction in compression
X_ε	Allowable strain in the longitudinal direction in compression
Y_ε	Allowable strain in the transversal direction in compression
S_ε	Shear strain limit
σ_{red}	Stresses reduced to uniaxial stresses
$\sigma_1, \sigma_2, \sigma_3$	Nominal stresses ($\sigma_1 > \sigma_2 > \sigma_3$)
$\varepsilon_1, \varepsilon_2, \varepsilon_3$	Nominal strains ($\varepsilon_1 > \varepsilon_2 > \varepsilon_3$)
E, E_i	Young Modulus, Young Modulus in i-th direction
G, G_{ij}	Kirchhoff Modulus, Kirchhoff modulus in the direction of ij plane normal
σ_{ij}	Elements of stress tensor
$\mathbb{I}$	Identity matrix
C^0, C^d	Undamaged and damaged stiffness matrix
D	Damage parameter, Damage parameter matrix
σ_0, σ_d	Stress tensor for undamaged material, stress tensor for damaged material
d_f	Damage of fiber level
d_m	Damage of matrix level
d_s	Shear damage level

3.1 Introduction

In the previous chapter mechanical properties of fiber-reinforced plastics are described. However, an equally important part of their assessment is a description of losing these properties. In the damage process, we distinguish two main stages:

- Damage initiation—which indicates moment, when the damage process starts to occur,
- Damage propagation—which describes how material loses its mechanical properties with further loading after damage initiation point.

In this chapter, we will also focus on the implementation of these two aspects of damage to numerical methods, as they are today integral tools in the analysis of materials and structures.

Fiber-reinforced composite materials can be analyzed with different levels of detail. Usually, we distinguish three approaches [1, 2] based on that factor:

- ***Micromechanical***—where we consider materials of fibers and matrix separately as well as include the investigation of the bond behavior between them. While this approach is the most realistic one it is also the most complex and is hard

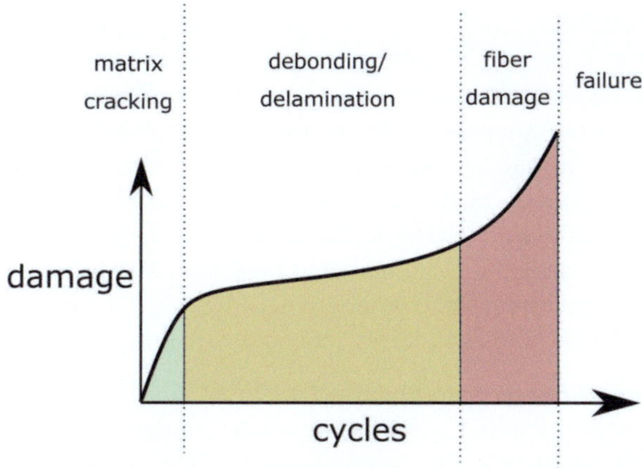

Fig. 3.1 Three stages of damage in unidirectional composites as proposed in 1983 by Reifsnider et al. based on [4]

to implement in models containing whole structures or even specimens. Usually, its usage is limited to the analysis of unit cells, Representative Volume Elements (RVE) or similar.

- ***Mesomechanical***—in the case of laminates in this level, we distinguish each layer separately as an anisotropic body—it is especially important when the layers have different fiber orientations. Similarly, this approach can be useful also for hybrid bars/rods when we have different material properties over the radius of the rod or we have additional over-wrap made using filament winding. In this approach cohesion of the material is assumed to be perfect or can be modeled using, e.g., cohesive zone model. A lot of models that describe behavior of material on this level will be phenomenological models, i.e., such that they will be not investigating the real microstructure and interface of constituents, but rather try to describe the layer as a whole.
- ***Macromechanical***—in the case of laminates this is an approach that uses homogenized material by following for example classical laminate theory or similar concepts. While it can be successful in the description of behavior for classic composites it can't be used for more complex/hybrid composites as it leads to significant error due to much simplification.

In 1983 Reifsnider [3] proposed three stages of damage in unidirectional composites. It was a good base for further analysis in that area, however, the issue is much more complex. Nevertheless, this proposition is presented below in Fig. 3.1.

3.1.1 Damage Initiation

In this sub-section, we will focus on mesomechanical approach if not indicated otherwise. First, we will focus on ***damage initiation***, as this is the first stage of damage assessment. The topic of damage initiation in composites should be started with limit criteria—they are the most simple idea in damage assessment. In their case the reasoning is as follows—every material has some limits defined with respect to different components of the stress tensor. These limits can be also defined according to strain components. Such limits can be found by carrying out experiments following specific loading conditions for each case. Then, we can analyze damage by comparing components of stress tensor or strains with these limits. If the limits are exceeded, we believe that damage occurs. Below in Table 3.1 summary of limit criteria for both stress and strain formulation is presented with reference to the assessed damage type.

Maximum strain criterium is widely used in fatigue problems, where an important assumption is that fibers are equally loaded.

Limit-based criteria (also known as non-interactive) can be reliable, but only in specific loading conditions. It can be compared to the evolution of the classic strength criteria. Galileo's hypothesis formulated in 1632 assumed that material will fail if the tensile stresses are larger than the allowable tension stress $\sigma_{\mathrm{red}} = \sigma_1 > X$. Small modifications were proposed by Clebsch (1862) and Rankin (1856) assuming that failure is based on extremal normal stresses:

$$\sigma_{\mathrm{red}} = \max\left(\sigma_1, -\sigma_3 \cdot \frac{X}{X'}\right) < X \tag{3.1}$$

These criteria can be used for some specific problems like brittle material splitting, but they can't be used for most engineering issues. This is why more modern criteria were introduced that include a combination of different parts of the stress tensor (for example hypothesis $\tau_{\max}$ by Coloumb, Tresca, and Guest or maximum distortion criterium by Huber, Mises, and Hencky. Similar to this idea in order to enhance the applicability of the limit-based criteria multiple solutions were proposed that assess failure based on some combination of the stress tensor elements. They are usually named ***interactive criteria***. Below the most popular criteria will be briefly discussed.

The base for strength assessment of orthotropic materials is a method proposed by Hill in 1948 in his work "A theory of the yielding and plastic flow of anisotropic metals" [5]. It was a modification of von Mises's proposition by adding some coefficients that were a result of anisotropy. By assuming all these coefficients equal to 1 one can get exactly the von Mises criterion. This criterion is usually now known under

Table 3.1 Limit criteria of damage initiation in composite material

Criterion	Condition	Damage type
Maximum stress	$\frac{\sigma_{11}}{X} \geq 1$	Fiber failure
	$\frac{\sigma_{22}}{Y} \geq 1$	Transverse matrix cracking
	$\frac{\sigma_{12}}{S} \geq 1$	Shear matrix cracking
Maximum strain	$\frac{\varepsilon_{11}}{X_\varepsilon} \geq 1$	Fiber failure
	$\frac{\varepsilon_{22}}{Y_\varepsilon} \geq 1$	Transverse matrix cracking
	$\left\lvert\frac{\varepsilon_{12}}{S_\varepsilon}\right\rvert \geq 1$	Shear matrix cracking

the name of Hill–Tsai because of the modification proposed by Tsai in "Strength Characteristics of Composite Materials" published in 1965 [6]. The widely used shape of this criterion for plane stress conditions is presented below in Eq. 3.1.

$$\frac{\sigma_{11}^2}{X^2} + \frac{\sigma_{22}^2}{Y^2} + \frac{2\sigma_{11}\sigma_{22}}{X^2} + \frac{\tau_{12}^2}{S^2} = 1 \tag{3.2}$$

While the Tsai-Hill criterion is widely used (also due to being present by default in commercial software like Abaqus) it has two significant flaws. First, it always takes all damage modes included into account even if they are not present in the given damage. Secondly, it does not include distinguishing between allowable stresses in tension and compression, which is often the case for composite materials. This issue can be easily addressed by modifying allowable stresses based on the real state of the stress. These changes were proposed by both Tsai and Azzi [7] and the final version of the criterion written below is often referred to as the Azzi-Tsai-Hill criterion.

$$\frac{\sigma_{11}^2}{X^{*2}} + \frac{\sigma_{22}^2}{Y^{*2}} + \frac{2\sigma_{11}\sigma_{22}}{X^{*2}} + \frac{\tau_{12}^2}{S^2} = 1 \tag{3.3}$$

$$X^* = \begin{cases} X \text{ if } \sigma_1 > 0 \\ X' \text{ if } \sigma_1 \leq 0 \end{cases}$$

Yet another variant was proposed by Tsai and Wu in their work "A General Theory of Strength for Anisotropic Materials" published in 1971 [8] This variant is a simpler version of the more general theory for failure in anisotropic materials proposed by Gol'denblat and Kapnov (see [9]), but it was developed independently. Gol'denblat and Kapnov's approach was defined with infinite series and hence was not very useful in practice.

$$F_1\sigma_{11} + F_2\sigma_{22} + F_{11}\sigma_{11}^2 + F_{22}\sigma_{22}^2 + 2F_{12}\sigma_{11}\sigma_{22} + F_{66}\tau_{12}^2 = 1 \tag{3.4}$$

$$F_1 = \frac{1}{X} + \frac{1}{X'}; F_2 = \frac{1}{Y} + \frac{1}{Y'}; F_{11} = -\frac{1}{XX'}; F_{22} = -\frac{1}{YY'}; F_{66} = \frac{1}{S^2} \tag{3.5}$$

F_{12} have to be determined by experimental testing. It is also possible to obtain an approximate value of F_{12} coefficient by using one of the following propositions (Eqs. 3.6 or 3.7):

$$F_{12} \cong -\frac{1}{2}\sqrt{F_{11} \cdot F_{22}} \tag{3.6}$$

$$F_{12} \cong -\frac{1}{2}\sqrt{\frac{1}{X \cdot X' \cdot Y \cdot Y'}} \tag{3.7}$$

The main limitation of the Tsai-Wu criterion is fiber compression failure assessment. This is justified by the fact that such type of damage is usually based on geometry flaws and cannot be well defined by the material property, which is an assumption for Tsai-Wu and similar phenomenological models.

Tsai and Wu's proposition is also an extension of the previously described criterion proposed by Tsai and Hill. The former can be simplified to the latter with the assumption that $X = X'$ and $Y = Y'$.

Another extension of the Tsai-Hill approach which included the difference between compression and tension was proposed by Hoffman in "The Brittle Strength of Orthotropic Materials" published in 1967 [10]. Instead of relying on only quadratic expressions of allowable stresses, it is proposed to add odd functions to the main criterion. The result is presented below

$$\begin{aligned} & C_1(\sigma_{22} - \sigma_{33})^2 + C_2(\sigma_{33} - \sigma_{11})^2 + C_3(\sigma_1 - \sigma_2)^2 + C_4\sigma_1 \\ & + C_5\sigma_{22} + C_6\sigma_{33} + C_7\tau_{23}^2 + C_8\tau_{13}^2 + C_9\tau_{12}^2 \geq 1 \end{aligned} \tag{3.8}$$

So in total 9 material parameters have to be defined in order to use this criterion. They can be derived from allowable stresses obtained from tensile uniaxial tests (X, Y, Z), compression uniaxial tests (X', Y', Z'), and shear tests (Q, R, S).

For transversely isotropic materials in case of a planar state of stress type of problem, it can be simplified to

$$\frac{\sigma_{11}^2}{X \cdot X'} - \frac{\sigma_{11}\sigma_{22}}{X \cdot X'} + \frac{\sigma_{22}^2}{Y \cdot Y'} - \frac{X - X'}{X \cdot X'}\sigma_{11} - \frac{Y - Y'}{Y \cdot Y'}\sigma_{22} + \left(\frac{\tau_{12}}{S}\right)^2 < 1 \tag{3.9}$$

which shows similarities to previously discussed criteria.

It should be underlined that all interactive criteria presented above are phenomenological models, which means that while they can have good agreement with experimental testing results, they are not based on the mechanics of the material itself, but rather are fitted to experimental results. Nevertheless, they can be valid approaches to analyze damage in composites.

Interactive criteria have one principled disadvantage—they just decide if the damage occurs or not by checking the condition based on the failure index function in the form of

$$f\left(\left[\sigma_{ij}\right]\right) \geq 1 \tag{3.10}$$

Thus, they are not able to answer what was the main cause of the damage. This can be overcome by using ***separate mode criteria***. These criteria have a shape of a set of functions $f_k\left(\left[\sigma_{ij}\right]\right)$ which are calculated separately for each k mode of the damage. Usually, the mode includes some combination of the following aspects:

- Fiber compressive failure,
- Fiber tensile failure,
- Matrix tensile cracking,

- Matrix compressive cracking,
- Interlaminar compressive failure,
- Interlaminar tensile failure,
- Delamination.

The most well-known separate mode criterium is Hashin. It was first proposed in 1973 in "A Fatigue Failure Criterion for Fiber Reinforced Materials" [11] and developed further in 1980 in "Failure Criteria for Unidirectional Fiber Composites" [12]. First, this approach was designed for two-dimensional plane stress conditions—it includes four modes of damage: fiber tensile failure (while $\sigma_{11} > 0$)

$$f_{ft} = \left(\frac{\sigma_{11}}{X}\right)^2 + \alpha\left(\frac{\tau_{12}}{S}\right)^2 \geq 1 \tag{3.11}$$

(above formula is based on the Hashin-Rotem proposition with $\alpha = 0$ and also on a further version by Hashin with $\alpha = 1$ m), fiber compression failure (while $\sigma_{11} < 0$)

$$f_{fc} = \left(\frac{\sigma_{11}}{X'}\right)^2 \geq 1 \tag{3.12}$$

matrix tensile cracking (while $\sigma_{22} > 0$)

$$f_{mt} = \left(\frac{\sigma_{22}}{Y}\right)^2 + \left(\frac{\tau_{12}}{S}\right)^2 \geq 1 \tag{3.13}$$

And finally, matrix compression cracking (while $\sigma_{22} < 0$)

$$f_{mc} = \frac{1}{4}\left(\frac{\sigma_{22}}{R}\right)^2 + \frac{1}{4} \cdot \frac{Y'^2\sigma_{22}}{S^2Y'} - \frac{\sigma_{22}}{Y'} \geq 1 \tag{3.14}$$

The proposition defined in the Hashin 2D criterion can be extended to a more general 3D criterion in the following manner:

fiber tensile failure (while $\sigma_{11} > 0$)

$$f_{ft} = \left(\frac{\sigma_{11}}{X}\right)^2 + \left(\frac{\tau_{12}}{S}\right)^2 + \left(\frac{\tau_{13}}{Q}\right)^2 \geq 1 \tag{3.15}$$

fiber compression failure (while $\sigma_{11} < 0$)

$$f_{fc} = \left(\frac{\sigma_{11}}{X'}\right)^2 \geq 1 \tag{3.16}$$

matrix tensile cracking (while $\sigma_{22} > 0$)

$$f_{mt} = \left(\frac{\sigma_{22} + \sigma_{33}}{Y}\right)^2 + \frac{1}{R^2}\left(\tau_{23}^2 - \sigma_{22}\sigma_{33}\right) + \left(\frac{\tau_{12}}{S}\right)^2 + \left(\frac{\tau_{13}}{Q}\right)^2 \geq 1 \tag{3.17}$$

And finally, matrix compression cracking (while $\sigma_{22} < 0$)

$$f_{mc} = \left(\frac{\sigma_{22} + \sigma_{33}}{2R}\right)^2 + \frac{\sigma_{22} + \sigma_{33}}{Y'}\left(\left(\frac{Y'}{2R}\right)^2 - 1\right) + \frac{1}{R^2}\left(\tau_{23}^2 - \sigma_{22}\sigma_{33}\right) + \left(\frac{\tau_{12}}{S}\right)^2 + \left(\frac{\tau_{13}}{Q}\right)^2 \geq 1 \quad (3.18)$$

Both versions presented above are very similar and the main difference is that σ_{33} is included in the three-dimensional version, which is more reflecting the real behavior of the material [13].

Another well-known separate mode criterion is Chang-Chang, which proposes another approach to matrix compressive failure:

$$f_{mc} = \left(\frac{\sigma_{22}}{S}\right)^2 + \left(\frac{\tau_{12}}{S}\right)^2 + \frac{1}{4} \cdot \frac{Y'\sigma_{22}}{S^2} - \frac{\sigma_{22}}{Y'} \geq 1 \quad (3.19)$$

The above version of Chang-Chang is based on [13], however, various versions of it are in use in different applications. The criterion itself was proposed in different forms in works by Chang and Chang in 1987 in "A Progressive Damage Model for Laminated Composites Containing Stress Concentrations" [14] as an improvement to previous works published in 1978 by Yamada and Sun [15] as well as Chang, Scott and Springer in 1985 [16]. It should be underlined that in application to commercial software like MSC often a combination of different versions of criteria is used. Exemplary analysis of development and choosing of such criteria can be found for example in "A study of failure criteria for fibrous composite material" by Paris and Jackson [17]

The next common criterion is this proposed by Puck. The criterion can be described for example in the following way based on Puck and Schurmann's "Failure analysis of FRP laminates by means of physically based phenomenological models" published in 1986 [18].

fiber tensile failure (while $\sigma_{11} > 0$)

$$\frac{1}{\varepsilon_{11}}\left(\varepsilon_1 + \frac{\nu_{12}}{E_{f1}} m_{f\sigma}\sigma_2\right) \geq 1 \quad (3.20)$$

fiber compression failure (while $\sigma_{11} < 0$)

$$\left|\frac{1}{\varepsilon_{1T}}\left(\varepsilon_1 + \frac{\nu_{f12}}{E_{f1}} m_{f\sigma}\sigma_2\right)\right| \geq 1 - (10\gamma_{21})^2 \quad (3.21)$$

matrix tensile cracking (while $\sigma_{22} > 0$)

$$\sqrt{\left(\frac{\tau_{21}}{s_{21}}\right)^2 + \left(1 - P_{\perp\parallel}^{(+)} Y_T/s_{21}\right)^2 (\sigma_2/Y_T)^2} + p_{\perp\parallel}^{(+)}\sigma_2/s_{21} \geq 1 - |\sigma_1/\sigma_{1D}| \quad (3.22)$$

and finally matrix compression cracking (while $\sigma_{22} < 0$) for 2D plane stress Mode B, $\theta_{f_P} = 0°$

$$\frac{1}{s_{21}}\left(\sqrt{\left(\tau_{21}^2 + p_{\perp\parallel}^{(-)}\sigma_2\right)^2 + P_{\perp}^{c}\mathrm{i}^3|\sigma_2|}\right) \geq 1 - \left|\frac{\sigma_1}{\sigma_{113}}\right| \tag{3.23}$$

and for mode C $\theta_{f_P} \neq 0°$,

$$\left[\left(\frac{\tau_{21}}{2(1 + p_{\perp\perp}^{(-)}S_{21}}\right)^2 + \left(\frac{\sigma_2}{y_C}\right)\right]\frac{Y_C}{(-\sigma_2)} \geq 1 - \left|\frac{\sigma_1}{\sigma_{1D}}\right| \tag{3.24}$$

The Puck criteria are often considered one of the most reliable. This opinion has some confirmation in conclusions from World Wide Failure exercises. However, the Puck criterion requires obtaining very specific material data, which is usually not easy to do.

In the 2000s new proposition by Camanho, Pinho, and Davila was established as LaRC. It was based on the concepts proposed by Hashin and the fracture plane idea developed by Puck and also used some other previously declared concepts. Up to this point current version is denoted as LaRC05 and is a result of World-Wide Failure Exercise (WWFE-II). This criterion considers effects like kinking of the fibers, failure of the matrix and for laminates also delamination.

Below in Fig. 3.2 bibliographic analysis of the usage of different criteria is presented based on the Scopus queries. The results should be treated more in a qualitative aspect—as the exact numbers are dependent on the syntax of the query. However, for all queries, the syntax was analogical (with substitution on "Criterion name":

> TITLE-ABS-KEY ((FRP OR "fiber reinforced composite*" OR "fiber reinforced plastic*") AND (damage OR failure) AND CRITERION NAME

3.1.2 Damage Propagation

Up to this point of this chapter initiation criteria were covered. However, the equally important aspect is ***damage propagation*** that starts after initiation. Damage propagation is usually realized by the assumption that the mechanical properties of a material are locally changed during damage propagation. These can be done by introducing a tensor of damage parameters D that change parameters for particular terms of the stiffness matrix C^0:

$$C^n = (\mathbb{I} - D) \cdot C^0 \tag{3.25}$$

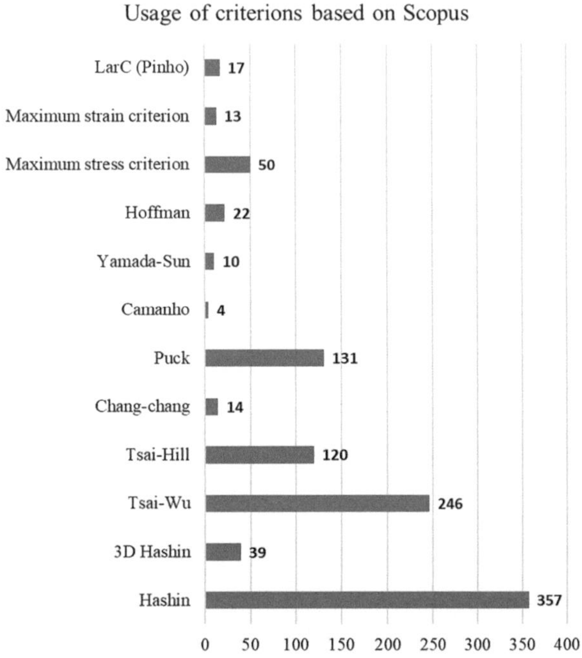

Fig. 3.2 Usage of various damage initiation criteria to investigate failure in FRP composites according to Scopus database queries

These approaches while generally intact require to use eight order damage tensor (as stiffness matrices are fourth-order tensors). Thus, simplification is proposed by making some assumptions toward the damage tensor. Here we will discuss shortly three most used ones based on a thesis by Crespo [19]:

- The strain equivalence principle—here we will assume that strains remain the same in the damaged material in comparison to undamaged ones. This led to the substitution of the stiffness tensor (fourth-order) for the stress tensor (second-order)

$$\sigma_d = (\mathbb{I} - D) \cdot \sigma_0 \tag{3.26}$$

- The stress equivalence principle—here we will assume that stress remains the same in the damaged material in comparison to the undamaged one. This led to the substitution of the stiffness tensor (fourth-order) for the strain tensor (second-order)

$$\varepsilon_d = (\mathbb{I} - D) \cdot \varepsilon_0 \tag{3.27}$$

- The energy equivalence principle—here we will assume that Helmholtz free energy Ψ remains the same in the damaged material in comparison to the undamaged one. This led to the substitution of the stiffness tensor (fourth-order) to strain tensor (second-order)

$$\sigma_d = (\mathbb{I} - D) \cdot C_0 \cdot (\mathbb{I} - D) \cdot \varepsilon \tag{3.28}$$

The simplest approach to introduce damage propagation is to use instant stiffness reduction (so it uses the concept of step function). In this approach, once the damage initiation criteria are reached the stiffness is reduced one-time to predefined values. This way, elements still carry some part of the stresses, but the response of the material is changed compared to the undamaged version. Of course, this approach is unrealistic, as the material rather loses its local stiffness gradually. Thus, a better description requires the introduction of so-called damage evolution laws which describe how the stiffness is lost after damage initiation occurs usually in terms of effective strains or displacement. Finally, after critical values of strain/displacement are reached, the cohesion of the material in such a place is lost. More information about approaches to damage evolution is described in the subchapter devoted to numerical implementations of damage prediction.

3.2 Cohesive Zone Model and R-Curve Effect

In the laminated composites besides mentioned above causes of damage there is also another important factor—which is delamination—loss of cohesion between layers. Methods that are in use for laminated composites can be also used for unidirectional FRP bars or rods to investigate their fracture behavior. To assess the delamination process in this type of material various tests are conducted reflecting particular modes of damage or mixed-mode:

- DCB (Double Cantilever Beam)—this is a test that reflects mode I type of loading—opening of the crack. It is realized on a pre-cracked specimen. In laminates, the precrack usually is obtained by introducing PTFE film between layers during the manufacturing process. In the case of pultruded bars or rods, the precrack is made by cutting the specimen.
- ENF (End Notch Flexural)—this is the test that reflects mode II type of loading—shearing the crack. It is realized on pre-cracked specimens which can be manufactured in a similar way as for the DCB test. The test is later realized using equipment designed for a 3-point bending test.
- MMB (Mixed Mode Bending)—this is the test that reflects mixed mode I + II type of loading—opening and shearing the crack. It is realized on pre-cracked specimens which can be manufactured in a similar way as for the DCB test. In Fig. 3.3 experimental stand-up equipped with additional apparatus is presented.

By adjusting geometrical parameters of the set-up like $2L$ (distance between supports and c (distance between force direction and central intender), it is possible to obtain different (but constant for one experiment) mixed-mode ratio. Value of c affects the forces acting for the opening of the crack (like in DCB) and shearing of the crack (like in ENF).

Fig. 3.3 Experimental stand-up with a universal testing machine equipped with apparatus for Mixed-mode bending test

Mentioned above experiments are useful for determining fracture parameters used in the cohesive zone model (CZM) theory. This theory is now widely used for modeling interfaces in composites but also can be used to model damage in rods or layers. The idea of a cohesive zone model is based on a general idea proposed in the 1960s by Dugdale [20] and Barenblatt [21]. Further development was realized by Willis [22] who used ideas from Dugdale and Barenblatt along with the fracture theory described by Griffith. The main idea of CZM is an assumption that fracturing of the material is localized around the crack tip and can be described by stresses σ in function of opening displacement δ. The heart of the method is traction–separation law (TSL) which describes this relationship:

$$\sigma = f(\delta) \tag{3.28}$$

In practice, different variants of function f are in use including the most popular ones: bilinear (triangular), trapezoid or exponential. Maximum value σ_c is often referred to as initiation criteria, and propagation is ruled by the shape of the curve and either area under the curve (fracture energy G_c) or critical displacements (δ_c). Depending on function f this model can include the following stages of development of damage:

- Elastic behavior (no damage yet),
- Plastic behavior,
- Damage evolution,
- Full debonding (no cohesion).

The exemplary traction–separation law is presented in Fig. 3.4.

Usually, fracture parameters for the cohesive zone model can be determined using experimental methods described previously in this chapter like DCB, ENF, or MMB with eventual numerical help. However, in some materials, the so-called "R-curve effect" can be present, which is indicated by the fact, that fracture energy required to

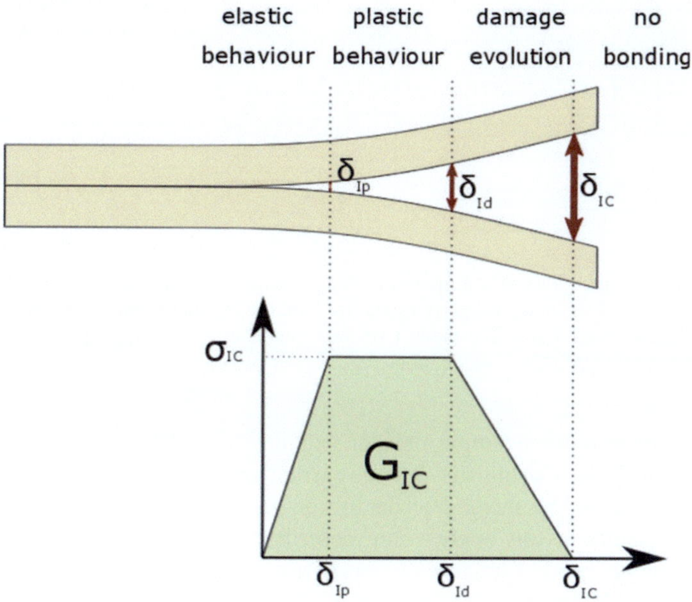

Fig. 3.4 Trapezoidal traction—separation law. Four stages of damage development are presented. (Graphic by Authors originally published in such form in A Review on Finite-Element Simulation of Fibre Metal Laminates [23] used under http://creativecommons.org/licenses/by/4.0/)

increase crack length is not constant but rather depends on the current crack length. The R-curve can be rising (fracture energy increases with longer cracks) or falling (fracture energy decreases with longer cracks). An exemplary R-curve is presented below in Fig. 3.5.

Fig. 3.5 Exemplary R-curve describing a change in fracture energy depending on the size of the defect (crack length). (Graphic by Authors originally published in such form in Combined experimental–numerical mode I fracture characterization of the pultruded composite bars [24] used under http://creativecommons.org/licenses/by/4.0/)

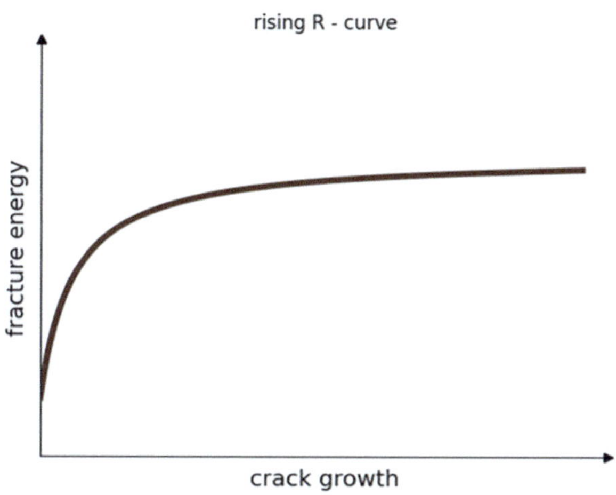

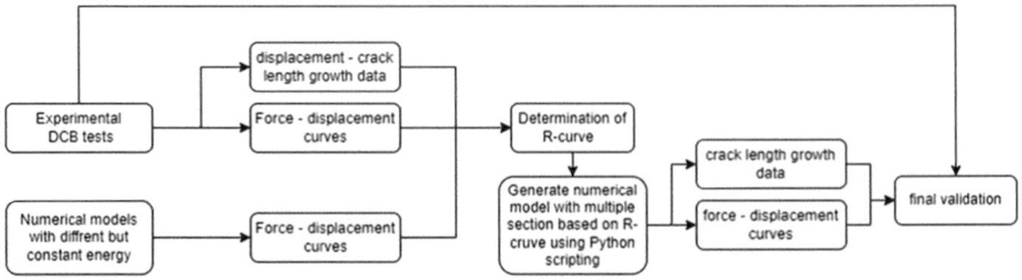

Fig. 3.6 The schematic idea of a proposed combined experimental–numerical method of determining R-curve for FRP composites. (Graphic by Authors originally published in such form in Combined experimental–numerical mode I fracture characterization of the pultruded composite bars [24] used under http://creativecommons.org/licenses/by/4.0/)

This behavior is for example present in pultruded FRP rods produced by the Authors' team [24]. During the project carried out on such materials combined experimental–numerical method of determining fracture energy and R-curve was developed. Results of this analysis are described in the paper Combined experimental–numerical mode I fracture characterization of the pultruded composite bars published in 2023. The block diagram of this method is presented in Fig. 3.6.

3.3 Damage Prediction in Numerical Simulations

Currently, the most popular software used for modeling the initiation and damage propagation in composite materials is Simulia Abaqus, Ansys, LS-Dyna. The popularity of the software was also checked by investigating the databases like Scopus. For the purpose of this chapter, we created a set of bibliographic queries with the following format:

TITLE-ABS-KEY ((FRP OR "fiber reinforced composite*" OR "fiber reinforced plastic*") AND (damage OR failure) AND SOFTARE_NAME)

In these queries, the TITLE-ABS-KEY means that all the papers containing declared words in the title, abstract, or keywords will be returned. SOFTWARE_NAME was substituted with various numerical software. The asterisk * is used as a string with any number of characters. Finally AND, OR are used naturally as logical connectives. The results of the analysis are presented in Fig. 3.7. Exact numbers are dependent on the phrasing of the queries, but the ratio is stable. Other software is not present in the database at all.

The most popular of them—Abaqus—has some of the models described in this chapter already implemented: limit-based criteria (maximum stress and strain) and interactive criteria can be used for both plane stress and general elements. However, with them, it is not possible to model further propagation, but only point to the moment of the damage initiation. Besides them, a separate mode criterium is implemented (Hashin or Hashin-Rotem), but it is limited to plane stress elements, which

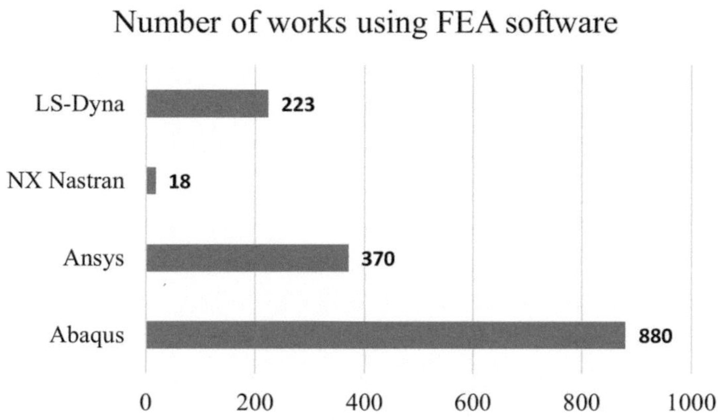

Fig. 3.7 Count of works that were using FEA software which is returned from a query to the Scopus database

include plane stress, shell, continuum shell, and membrane elements. Fortunately, a three-dimensional version of this criterium can be implemented independently by researchers by using a user material subroutine (UMAT or VUMAT) written in FORTRAN, which is a widely used option (e.g., [25–27]).

Ansys is the second most popular software to deal with composite damage utilizing the finite element method. Some of the most popular criteria are predefined to use. These include maximum stress, Tsai-Wu, Tsai-Hill, LaRC, Cuntze, and others. Similar to Abaqus it is possible to use other criteria after self-implementation using APDL (Ansys Parametric Design Language). This option is often used to extend the possibilities of a plain program.

LS-Dyna is traditionally recognized as software focused on dynamic problems, but can be used for a wide variety of tasks. There are a few material models used in LS-Dyna for dealing with composite damage. Material models 22, 54, and 55 are general models for composite damage that follow mostly Chang-Chang criteria (Tsai-Wu for matrix in model 55). Material models 58, 59, and 158 follow the modified Hashin criterion. There are also two less typical models available—material model 261 follows the Pinho criterion [28, 29] (with Puck criterion for inter-fiber failure) and model 262 follows the Camanho criterion [30, 31] (with Puck criterion for inter-fiber failure). In LS-Dyna users have the possibility to define their material using FORTRAN similar to Simulia Abaqus.

In Table 3.2, the summary of predefined damage initiation criteria in the most popular FEA software is presented. With "x" available criteria are indicated for each software.

In the case of Simulia Abaqus progressive damage is realized with the assumption that undamaged composites behave in a linear-elastic manner and the failure is mostly brittle (with small plastic deformation). The propagation of the damage is calculated using dissipated energy and reduction of the stiffness that can be defined in various

Table 3.2 Summary of predefined damage initiation criteria in the most popular FEA software. Criteria are separated by their type

	Limit based criteria		Interactive criteria				Separate mode criteria							
	Maximum stress	Maximum strair	Tsai-Wu	Tasi-Hill	Azzi-Tasi-Hill	Hoffman	Hashin-Rotem	Hashin 2D	Hashin 3D	Chang-Chang	LaRCOx (Pinho)	Cuntze	Camanho	Puck
Abaqus	X	X	X	X	X		X	X			X			
Ansys	X		X	X				X			X	X		X
LS-Dyna			X					X		X	X		X	X
MSC Patran														
Captia	X		X	X		X								
NX Nastran		X	X	X		X								

ways. Abaqus documentation [32] declares that the failure mechanism is defined in four steps:

- Undamaged material response—which is assumed to be linear-elastic—which is close to the reality for FRP materials,
- Damage initiation criterion (as described in detail above),
- Damage initiation law—it is introduced by affecting the elasticity matrix C_d. The general shape of the elasticity matrix for damaged materials is presented below

$$C_d = \frac{1}{D}\begin{bmatrix} (1-d_f)E_1 & (1-d_f)(1-d_m)\nu_{21}E_1 & 0 \\ (1-d_f)(1-d_m)\nu_{12}E_2 & (1-d_m)E_2 & 0 \\ 0 & 0 & (1-d_s)GD \end{bmatrix} \quad (3.29)$$

where

$$D = 1 - (1-d_f)(1-d_m)\nu_{12}\nu_{21}$$

d_f—damage of fiber level, d_m—damage of matrix level, d_s—shear damage level.

It should be underlined that such damage evolution can only be used with the Hashin criterion or LaRC05. Damage evolution is assessed separately for each of four main modes of damage:

- Fiber rupture in tension;
- Fiber buckling and kinking in compression;
- Matrix cracking under transverse tension and shearing; and
- Matrix crushing under transverse compression and shearing.

Softening in Simulia Abaqus can be defined in three different ways: linear, exponential and user-defined values in the form of a table. This gives a variety of options during the definition of material behavior.

- The decision about element deletion after critical strain is reached

In the ANSYS there are two implemented possibilities of damage evolution: instant stiffness reduction (MPDG) and continuum damage mechanics (PDA). The former can be used with any of the damage initiation criteria available in ANSYS (like Puck, LarC, etc.) but it does not reflect the damage evolution, as the reduction in stiffness is a one-time behavior at the damage initiation. On the other hand, PDA can be used only with the Hashin criterion but enables gradual change of stiffness [33]. The damage evolution law in ANSYS is bound to be bilinear. It is defined separately for each of the four damage modes by declaring for each energy dissipated per unit area G and viscosity η.

The importance of user-defined material models by UMAT/VUMAT or APDL can be underlined by investigating other similar queries to the Scopus database, e.g.:

> FRP OR "fiber reinforced composite*" OR "fiber reinforced plastic*" AND damage OR failure AND Abaqus AND (UMAT) OR (VUMAT)

These queries return 135 results for Abaqus UMAT/VUMAT and another 15 for Ansys APDL. The presence of such tools makes it possible to check in an easy way effects and differences for newly defined damage initiation criteria or damage propagation schemes.

An important aspect of predicting damage with an analytical approach or numerical simulations is to verify its quality.

By using various non-destructive methods it is possible to check what happened during the loading of the material and assigned material response to particular types of damage that include matrix cracking, debonding, fiber breaking and delamination. Based on work by Duchene et al. "A review of non-destructive techniques used for mechanical damage assessment in polymer composites" [34] range of applications of different techniques is considered:

- Acoustic emission (microscale to macroscale)
- X-ray tomography (microscale to mesoscale)
- X-ray radiography (mesoscale to macroscale)
- Ultrasonic waves (mesoscale to macroscale)
- Infrared thermography (mesoscale to macroscale)
- DIC (macroscale + delamination in mesoscale).

Additionally, destructive methods mostly connected to microscopy (light, scanning electron, transmission) can be also used for similar purposes.

3.4 Summary

With the constant development of composite applications to structures, objects and buildings there is a need to analyze failure behavior and predict damage or lifetime. From the 1970s many research groups work on this topic proposing different approaches to damage initiation and damage propagation. Currently, the most used criteria are based on the works of Hashin combined with gradual damage propagation. Nowadays these theory bases are used commonly with finite element modeling which enables us to use them on the macroscale of construction and buildings. They are either predefined in the most popular software like Simulia Abaqus, Ansys, or LS-Dyna or can be introduced by writing their code in the shape of subroutines. Development of new composites continues—looking for new types of fibers, matrices or manufacturing technologies. Thus, there is still room for new, tailored criteria of failure, which can be better in some cases than currently used ones. Also, it is possible to investigate aspects such as uneven distribution of fibers, voids, manufacturing issues or property heterogeneity. Other widely recognized approaches such as the cohesive zone model can be also used in the damage prediction of composites. However, some unusual effects can be spotted like the "R-curve effect" so modern combined approaches are developed to be able to determine all needed parameters.

Damage prediction has a crucial meaning in the assessment of composites as it enables better design of FRP materials taking into account potential applications and current state-of-the-art.

Acknowledgements The research was financed by the National Center for Research and Development of Poland (NCBiR) grant number LIDER/40/0219/L-10/18/NCBR/2019.

References

1. Soltani P, Keikhosravy M, Oskouei RH, Soutis C (2011) Studying the tensile behaviour of GLARE laminates: a finite element modelling approach. Appl Compos Mater 18:271–282. https://doi.org/10.1007/s10443-010-9155-x
2. Smolnicki M, Stabla P (2019) Finite element method analysis of fibre-metal laminates considering different approaches to material model. SN Appl Sci 1:467. https://doi.org/10.1007/s42452-019-0496-2
3. Reifsnider KL, Schulte K, Duke JC (1983) Long-term fatigue behavior of composite materials. In: ASTM Special Technical Publication
4. Ansari MTA, Singh KK, Azam MS (2018) Fatigue damage analysis of fiber-reinforced polymer composites—a review. 37:636–654. https://doi.org/10.1177/0731684418754713
5. Hill R (1948) A theory of the yielding and plastic flow of anisotropic metals. Proc R Soc Lond A Math Phys Sci 193:281–297. https://doi.org/10.1098/RSPA.1948.0045
6. Tsai S (1965) Strength characteristics of composite materials
7. Azzi VD, Tsai SW (1965) Anisotropic strength of composites - investigation aimed at developing a theory applicable to laminated as well as unidirectional composites, employing simple material properties derived from unidirectional specimens alone. Exp Mech 5:283–288. https://doi.org/10.1007/BF02326292/METRICS
8. Tsai SW, Wu EM (1971) A general theory of strength for anisotropic materials. J Compos Mater 5. https://doi.org/10.1177/002199837100500106
9. Gol'denblat II, Kopnov VA (1984) Experimental verification of strength criteria for anisotropic materials. Strength Mater 16:403–405. https://doi.org/10.1007/BF01530038/METRICS
10. Hoffman O (1967) The Brittle strength of orthotropic materials. J Compos Mater 1. https://doi.org/10.1177/002199836700100210
11. Hashin Z, Rotem A (1973) A fatigue failure criterion for fiber reinforced materials. J Compos Mater 7:448–464. https://doi.org/10.1177/002199837300700404
12. Hashin Z (1980) Failure criteria for unidirectional fiber composites. J Appl Mech 47:329–334. https://doi.org/10.1115/1.3153664
13. Shao JR, Liu N, Zheng ZJ (2021) Numerical comparison between Hashin and Chang-Chang failure criteria in terms of inter-laminar damage behavior of laminated composite. Mater Res Express 8:085602. https://doi.org/10.1088/2053-1591/AC1D40
14. Chang F-K, Chang K-Y (1987) A progressive damage model for laminated composites containing stress concentrations. J Compos Mater 21:834–855. https://doi.org/10.1177/002199838702100904
15. Yamada SE, Sun CT (1978) Analysis of laminate strength and its distribution. J Compos Mater 12. https://doi.org/10.1177/002199837801200305
16. Chang FK, Scott RA, Springer GS (1984) Failure strength of nonlinearly elastic composite laminates containing a pin loaded hole. J Compos Mater 18. https://doi.org/10.1177/002199838401800506
17. Paris F, Jackson KE (2001) A study of failure criteria of fibrous composite materials. NASA reports

18. Puck A, Schürmann H (1998) Failure analysis of FRP laminates by means of physically based phenomenological models. Compos Sci Technol 58:1045–1067. https://doi.org/10.1016/S0266-3538(96)00140-6
19. Inês Maria Trindade Crespo (2015) Damage propagation in composite materials meso-mechanical models. Universidade Nova de Lisboa
20. Dugdale DS (1960) Yielding of steel sheets containing slits. J Mech Phys Solids 8:100–104. https://doi.org/10.1016/0022-5096(60)90013-2
21. Barenblatt GI (1962) The mathematical theory of equilibrium cracks in brittle fracture. Adv Appl Mech. https://doi.org/10.1016/S0065-2156(08)70121-2
22. Willis JR (1967) A comparison of the fracture criteria of griffith and barenblatt. J Mech Phys Solids. https://doi.org/10.1016/0022-5096(67)90029-4
23. Smolnicki M, Lesiuk G, Duda S, de Jesus AMP (2022) A review on finite-element simulation of fibre metal laminates. Arch Comput Methods Eng 1:1–15. https://doi.org/10.1007/S11831-022-09814-8/FIGURES/5
24. Smolnicki M, Duda S, Zielonka P et al (2023) Combined experimental–numerical mode I fracture characterization of the pultruded composite bars. Arch Civil Mech Eng 23:1–11. https://doi.org/10.1007/S43452-023-00684-W
25. Arruda MRT, Trombini M, Pagani A (2023) Implicit to explicit algorithm for ABAQUS standard user-subroutine UMAT for a 3D Hashin-based orthotropic damage model. Appl Sci 13:1155. https://doi.org/10.3390/APP13021155
26. Chi Pham D, Cui X, Ren X, Lua J (2019) A discrete crack informed 3D continuum damage model and its application for delamination migration in composite laminates. https://doi.org/10.1016/j.compositesb.2019.02.045
27. Wang G-D, Kirwa Melly S (2018) Three-dimensional finite element modeling of drilling CFRP composites using Abaqus/CAE: a review. https://doi.org/10.1007/s00170-017-0754-7
28. Pinho ST, Iannucci L, Robinson P (2006) Physically based failure models and criteria for laminated fibre-reinforced composites with emphasis on fibre kinking. Part II: FE implementation. https://doi.org/10.1016/j.compositesa.2005.06.008
29. Pinho ST, Iannucci L, Robinson P (2006) Physically-based failure models and criteria for laminated fibre-reinforced composites with emphasis on fibre kinking: Part I: Development. https://doi.org/10.1016/j.compositesa.2005.04.016
30. Maimí P, Camanho PP, Mayugo JA, Dávila CG (2007) A continuum damage model for composite laminates: Part I - Constitutive model. Mech Mater 39:897–908. https://doi.org/10.1016/J.MECHMAT.2007.03.005
31. Maimí P, Camanho PP, Mayugo JA, Dávila CG (2007) A continuum damage model for composite laminates: Part II – Computational implementation and validation. Mech Mater 39:909–919. https://doi.org/10.1016/J.MECHMAT.2007.03.006
32. Abaqus 6.13 (2013) Abaqus 6.13. Analysis user's guide, Dassault Systems
33. Barbero EJ, Shahbazi M (2017) Determination of material properties for ANSYS progressive damage analysis of laminated composites. Compos Struct 176:768–779. https://doi.org/10.1016/J.COMPSTRUCT.2017.05.074
34. Duchene P, Chaki S, Ayadi A, Krawczak P (2018) A review of non-destructive techniques used for mechanical damage assessment in polymer composites. J Mater Sci 53:7915–7938. https://doi.org/10.1007/S10853-018-2045-6

Chapter 4
Resistance to Harsh Environment

**Adrianna Filipiak-Kaczmarek, Michał Barcikowski,
and Wojciech Błażejewski**

Abstract Reinforcement bars in many cases are subjected to harsh environment besides the mechanical load. This environment has significant impact if offshore structures are considered, however, the concrete itself may affect the reinforcing bars. In terms of metallic materials those degradation mechanisms are well-described, on the other hand if composite materials are under consideration, the process is more complex and still requires investigation. In this chapter, research of the harsh environment on flexural behavior is described and quantified. Two different solutions were used to simulate harsh environment, e.g., alkaline and salty water, the research is conducted in elevated temperature.

Keywords Harsh environment · Flexural behavior · Degradation mechanisms

4.1 Introduction

The durability of FRP rods is an important issue from the point of view of the strength of concrete structures that are reinforced with it. The properties of the bars are affected by many factors, both external and internal: the type of matrix type of fibers, the fibers/matrix interface, exposure to various solutions (alkaline, acidic, alkaline, etc.), exposure to water and different temperature ranges.

Aging testes of FRP reinforcing bars are conducted in accordance with ASTM standards. The most popular simulated environments are alkaline, salty or acidic and high pH environments. However, choice of environments isn't random. FRP reinforcing bars are generally used in concrete structures (alkaline solution) working in aqueous environments—most often marine (saline solution). Except for the choice of environmental in which the aging tests will be carried out, the selection of the

A. Filipiak-Kaczmarek (✉) · M. Barcikowski · W. Błażejewski
Faculty of Mechanical Engineering, Wroclaw University of Science and Technology, Wroclaw, Poland
e-mail: adrianna.filipiak@pwr.edu.pl

G. Lesiuk et al. (eds.), *Composite Rebars for Reinforced Concrete Structures*, Structural Integrity 34, https://doi.org/10.1007/978-3-031-96292-9_4

time interval is very important. It is assumed that 21 days of immersion in an aggressive environment at 60 °C correspond to 14 years of actual use, whereas 42 days correspond to 28 years.

The effect of aggressive environments on the strength of FRP bars is clear. In many publications it has been observed that with longer immersion time there is a decrease in strength parameters, which is also been confirmed by many test results.

The aging tests are very important—they are good introduction to further research and enable to checking several material parameters like measuring the weight before and after immersion, water absorption and measuring the pH of solution. Measuring the mass of samples is the first step to determine how is the impact aggressive solution on durability FRP rods. The decrease of mass a test to degradation of the resin, the resin-fiber interface or corrosion of the fiber itself. On the other hand, potential increase of mass gives information that the immersed samples have absorber the simulated environment. The mass loss/increase is also related to the absorption of water.

4.2 Experimental Program

4.2.1 Materials

Three types of FRP rods were investigated using accelerated aging tests, structural and mechanical tests. All FRP rods were made using the pultrusion method, Px250T3 pultrusion machine from PultreX was used for this purpose. Two types of fibers were used to produce the composite rebars: glass type E Hybon 2002 from PPG Fiber Glass Europe and basalt BCF 17-4800-KV42 from Kamenny Vek. Biresin CR144 epoxy resin was used as the matrix.

In this study were used three types of rods: glass rods with two diameters of 6 and 10 mm, basalt rods with a diameter of 6 mm, and hybrid rods with a diameter of 8 mm. The surface of the composite rebars was smooth. In addition the GFRP rods were divided into two groups: the first group was immersed under load and the second group without load. The hybrid rods were made of basalt fibers, which were placed in the central part of the rod, the so-called core, while glass fibers formed an envelope for the basalt core (Fig. 4.1).

4.2.2 Accelerated Aging

Two aging solutions were used for accelerated conditioning: alkaline and salty environments. The temperature of accelerated aging tests was 60 °C based on the conducted literature review. Forty specimens were exposed for 30 and 90 days. These periods would correspond to 20 and 60 years in a real concrete structure. After each

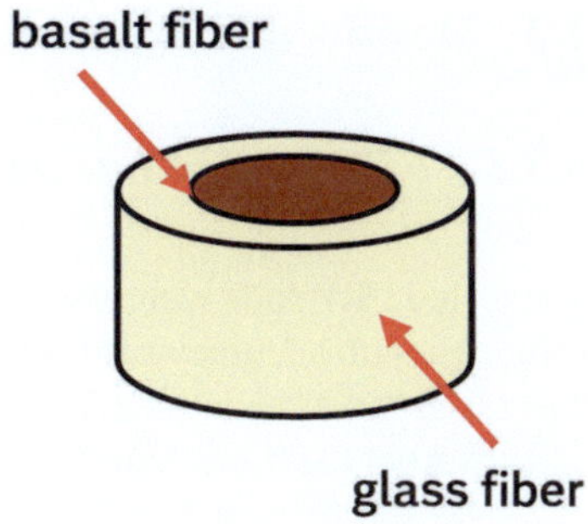

Fig. 4.1 Example of hybrid FRP rod construction

of these time intervals, 5 FRP rods of each kind (BFRP, GFRP, and HFRP) were pulled out of the aging chamber for structural and mechanical tests.

4.2.3 Determination of Reinforcement Content

The first step of the tests conducted was to weigh the melting pot themselves (to the nearest 0.001 g). Samples were inserted into the weighed crucibles and such a set was also weighed (to the nearest 0.001 g). The melting pot together with the samples were then placed in an oven heated to 650 °C. The calcination time was 6 h (± 1 h). After the samples were removed from the furnace, the composite reinforcement bars themselves were obtained. The melting pot set with the sample was weighed (to the nearest 0.001 g). Five samples of 20 mm in length from each FRP rod were used in the test.

4.2.4 Strength Tests—Four-Point Bending

In this study, a four-point bending test was conducted in accordance with ASTM 7264. Five samples of 300 mm in length from each FRP rod were used in the test. Totally, 80 specimens was tested. In addition, an extensometer was used during the trial to record the deflection of the specimens.

4.3 Results and Discussions

4.3.1 FRP Rods Characterization

The tested FRP rods were weighed after 30 and 90 days of immersion in aggressive solutions with accuracy of 0.001 g. For 67.5% of the samples, an increase in weight was observed after immersion in the environments, clearly indicating that the rods had absorber the fluids simulating the aggressive environments. However, a decrease in weight was recorded for 32.5% of the samples, which may be related to degradation at the resin-fiber interface and corrosion of the fiber itself, i.e., leaching of oxides from the fiber (Tables 4.1 and 4.2). The difference in the weight of the FRP rods before and after immersion in solutions was determined from the formula, where m_1 is a weight of specimen before immersion [g] and m_2 is a weight of specimen after immersion [g].

$$\Delta m = \frac{m_2 - m_1}{m_1} \cdot 100\%$$

The water absorption is quite important test, as it can determine the rate at which a material absorbs water after immersion. Combined with the determination of porosity and the change in the weight of the samples, water absorption can provide information that will help assessment the degradation of FRP bars.

In the group of samples immersed for 30 days in an alkaline environment, the most showed that both glass (GFRP10_O) and basalt (BFRP) samples showed an increase in mass after the 96-hour test period. For the other two types of samples, the trend was the opposite with a decrease in mass after the test period. However, for samples that were immersed in a saline solution an increase in mass was observed for basalt (BFRP) and hybrid (HFRP) rods. For glass rods (GFRP6) an increase in mass was recorded up to 48 h of immersion, while there was a decrease thereafter. Also for this pair of glass specimens, loaded (GFRP6_O) and unloaded (GFRP6_BO), statistical analyses showed the presence of statistically significant differences (Figs. 4.2 and 4.3).

Samples that were immersed in a saline environment for 90 days were characterized by an increase in weight after 96 h of water absorption. Similar results were obtained for samples immersed in an alkaline environment, where only glass samples were incubated under load (GFRP10_O) showed a decreasing trend. Statistically significant differences were also observed for glass rods (GFRP10) (Figs. 4.4 and 4.5).

However, comparing the results in the following groups: GFRP incubated with load for 30 and 90 days. GFRP incubated without load for 30 and 90 days and GFRP incubated with and without loaded for 90 days, it was observed that there is a correlation between water absorption and porosity. Statistically significant differences were noted in each of the aforementioned test groups in terms of the void content of the

Table 4.1 Summary of GFRP10 and GFRP6 specimens. Weight before (m_1) and after (m_2) incubation: 1M—30 days of incubation; 3M—90 days of incubation; No—incubation without load; Yes—incubation with load

Specimen	Solution	Immersion time	Load	m_1 [g]	m_2 [g]	Δm [%]
GFRP10_1	Alkaline	1M	Yes	49.831	49.824	−0.014
GFRP10_2	Alkaline	1M	Yes	47.279	47.224	−0.116
GFRP10_3	Alkaline	1M	Yes	47.542	47.423	−0.250
GFRP10_4	Alkaline	1M	Yes	48.846	48.815	−0.063
GFRP10_5	Alkaline	1M	Yes	49.542	49.448	−0.190
GFRP10_6	Alkaline	3M	Yes	50.205	50.121	−0.167
GFRP10_7	Alkaline	3M	Yes	46.776	46.756	−0.043
GFRP10_8	Alkaline	3M	Yes	50.408	50.341	−0.133
GFRP10_9	Alkaline	3M	Yes	49.143	49.021	−0.248
GFRP10_10	Alkaline	3M	Yes	48.812	48.706	−0.217
GFRP10_11	Alkaline	1M	No	45.429	45.436	0.015
GFRP10_12	Alkaline	1M	No	50.212	50.19	−0.044
GFRP10_13	Alkaline	1M	No	50.249	50.126	−0.245
GFRP10_14	Alkaline	1M	No	48.17	48.055	−0.239
GFRP10_15	Alkaline	1M	No	48.125	48.115	−0.021
GFRP10_16	Alkaline	3M	No	46.758	46.963	0.438
GFRP10_17	Alkaline	3M	No	47.365	47.453	0.186
GFRP10_18	Alkaline	3M	No	48.49	48.65	0.330
GFRP10_19	Alkaline	3M	No	48.803	48.879	0.156
GFRP10_20	Alkaline	3M	No	46.831	46.961	0.278
GFRP6_1	Saline	1M	No	16.953	17.015	0.366
GFRP6_2	Saline	1M	No	17.129	17.184	0.321
GFRP6_3	Saline	1M	No	16.774	16.821	0.280
GFRP6_4	Saline	1M	No	17.071	17.126	0.322
GFRP6_5	Saline	1M	No	16.99	17.041	0.300
GFRP6_6	Saline	3M	No	16.947	17.039	0.543
GFRP6_7	Saline	3M	No	16.734	16.838	0.621
GFRP6_8	Saline	3M	No	16.864	16.964	0.593
GFRP6_9	Saline	3M	No	17.027	17.123	0.564
GFRP6_10	Saline	3M	No	16.902	17.017	0.680
GFRP6_11	Saline	1M	Yes	17.134	17.199	0.379
GFRP6_12	Saline	1M	Yes	16.924	16.981	0.337
GFRP6_13	Saline	1M	Yes	16.854	16.919	0.386
GFRP6_14	Saline	1M	Yes	17.063	17.125	0.363
GFRP6_15	Saline	1M	Yes	17.135	17.201	0.385

(continued)

Table 4.1 (continued)

Specimen	Solution	Immersion time	Load	m_1 [g]	m_2 [g]	Δm [%]
GFRP6_16	Saline	3M	Yes	16.937	17.079	0.838
GFRP6_17	Saline	3M	Yes	17.264	17.428	0.950
GFRP6_18	Saline	3M	Yes	17.065	17.223	0.926
GFRP6_19	Saline	3M	Yes	17.178	17.316	0.803
GFRP6_20	Saline	3M	Yes	16.904	17.024	0.710

samples. In addition, it was observed that after immersing the rods in an alkaline environment, they were characterized by an almost 80% increase in weight.

4.3.2 Determination of Reinforcement Content

The content of reinforcement is important for the durability of FRP rods. This parameter determining how much reinforcement in a given composite is possible in several ways. The results of the study show that in each of the types of rods: glass, basalt and hybrid. The percentage content of fibers is at a high level and exceeds 75%.

For samples incubated in alkaline environment, it was observed that the lowest reinforcement content was in GFRP10 that were included without load. In addition, comparing the reinforcement content after 30 and 90 days, no change value was noted (Figs. 4.6 and 4.7).

For samples incubated in saline environment, a lower reinforcement content was observed than for samples from the previous group. Comparing the values in each group, no major changes in the reinforcement content were observed with longer immersion times. As in the case of FRP rods incubated in alkaline environment, also in this case statistical analyses showed no statistically significant differences (Figs. 4.8 and 4.9).

Determination of the reinforcement content was aimed at testing whether incubation of the rods in aggressive environments affects its reduction. The calcination test carried out made it possible to determine how the volume proportion of fibers in the composite changed. The results obtained clearly show that both the incubation time, the environment and whether the FRP rods were immersed with or without loading do not affect this parameter.

4.3.3 Four-Point Bending

The four-point bending test was conducted in accordance with ASTM. During the test. FRP rods that had previously been incubated in aggressive environments both with and without load were tested. The results obtained were analyzed taking into

Table 4.2 Summary of HFRP and BFRP specimens. Weight before (m_1) and after (m_2) incubation: 1M—30 days of incubation; 3M—90 days of incubation; No—incubation without load; Yes—incubation with load

Specimen	Solution	Immersion time	Load	m_1 [g]	m_2 [g]	Δm [%]
HFRP_1	Saline	1M	No	31.958	31.981	0.072
HFRP_2	Saline	1M	No	31.325	31.373	0.153
HFRP _3	Saline	1M	No	31.436	31.871	1.384
HFRP_4	Saline	1M	No	31.432	31.477	0.143
HFRP_5	Saline	1M	No	31.883	32.204	1.007
HFRP_6	Saline	3M	No	32.015	32.096	0.253
HFRP_7	Saline	3M	No	31.584	31.669	0.269
HFRP_8	Saline	3M	No	31.455	31.535	0.254
HFRP_9	Saline	3M	No	31.277	31.372	0.304
HFRP_10	Saline	3M	No	31.375	31.437	0.198
HFRP_11	Alkaline	1M	No	31.581	31.563	−0.057
HFRP_12	Alkaline	1M	No	31.442	31.396	−0.146
HFRP_13	Alkaline	1M	No	31.655	31.645	−0.032
HFRP_14	Alkaline	1M	No	31.822	31.74	−0.258
HFRP_15	Alkaline	1M	No	31.883	31.988	0.329
HFRP_16	Alkaline	3M	No	32.013	32.01	−0.009
HFRP_17	Alkaline	3M	No	32.152	32.794	1.997
HFRP_18	Alkaline	3M	No	31.844	32.106	0.823
HFRP_19	Alkaline	3M	No	31.461	31.503	0.133
HFRP_20	Alkaline	3M	No	31.476	32.2	2.300
BFRP_1	Saline	1M	No	17.873	17.936	0.352
BFRP_2	Saline	1M	No	17.918	17.973	0.307
BFRP_3	Saline	1M	No	17.835	17.894	0.331
BFRP_4	Saline	1M	No	18.068	18.126	0.321
BFRP_5	Saline	1M	No	17.821	17.888	0.376
BFRP_6	Saline	3M	No	18.038	18.24	1.120
BFRP_7	Saline	3M	No	17.942	18.059	0.652
BFRP_8	Saline	3M	No	17.724	17.897	0.976
BFRP_9	Saline	3M	No	17.918	18.007	0.497
BFRP_10	Saline	3M	No	17.733	17.828	0.536
BFRP_11	Alkaline	1M	No	17.96	17.987	0.150
BFRP_12	Alkaline	1M	No	17.881	17.947	0.369
BFRP_13	Alkaline	1M	No	17.935	17.969	0.190
BFRP_14	Alkaline	1M	No	17.843	17.868	0.140
BFRP_15	Alkaline	1M	No	17.921	17.946	0.140

(continued)

Table 4.2 (continued)

Specimen	Solution	Immersion time	Load	m_1 [g]	m_2 [g]	Δm [%]
BFRP_16	Alkaline	3M	No	17.913	17.761	−0.849
BFRP_17	Alkaline	3M	No	17.843	17.813	−0.168
BFRP_18	Alkaline	3M	No	17.85	17.782	−0.381
BFRP_19	Alkaline	3M	No	17.954	17.804	−0.835
BFRP_20	Alkaline	3M	No	17.937	18.117	1.004

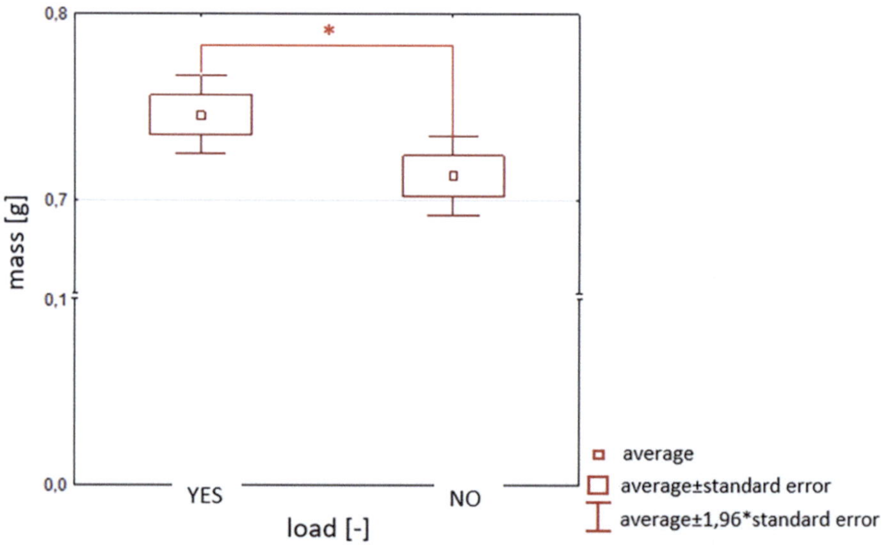

Fig. 4.2 Comparison of the weight of GFRP6_1M_O and GFRP6_1M_BO after 24 h of water absorption: Yes—samples incubated with load. No—samples incubated without unloaded; *—statistically significant differences ($p < 0.05$)

account the effect of time and the method of incubation, while the parameters that were determined were bending strength, strain, and Young's modulus. Statistical analyses showed statistically significant differences for several groups.

Basalt rods that were incubated in both alkaline and salty environments were characterized by a decrease in tensile strength with longer immersion time. As for the value of Young's modulus a decrease in its value was observed only for FRP rods from the saline environment, while an increase was recorded for those from the alkaline solution (Table 4.3).

For FRP bars that were immersed in alkaline solution statistical analyses showed statistically significant differences in the values of flexural strength and strain (Figs. 4.10 and 4.11). These results clearly indicate that immersion time has a significant effect on the strength of FRP rods.

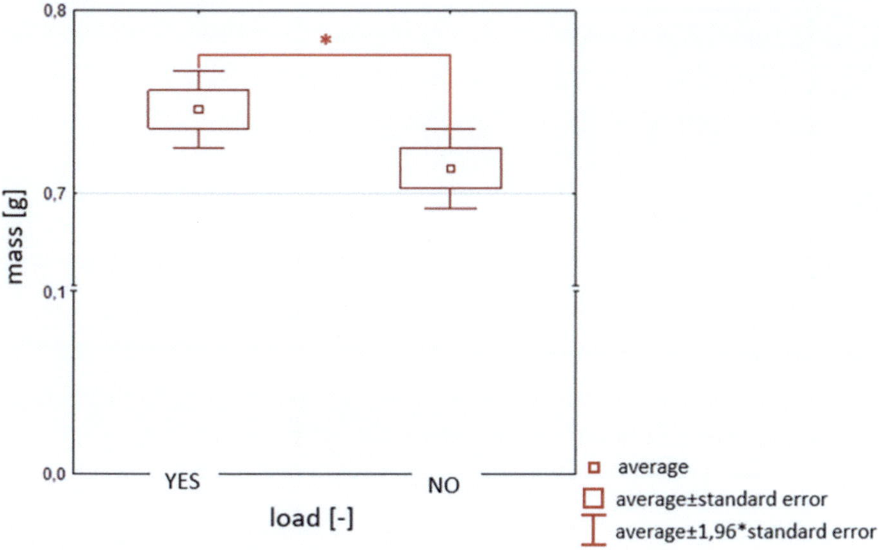

Fig. 4.3 Comparison of the weight of GFRP6_1M_O and GFRP6_1M_BO after 48 h of water absorption: Yes—samples incubated with load. No—samples incubated without unloaded; *—statistically significant differences ($p < 0.05$)

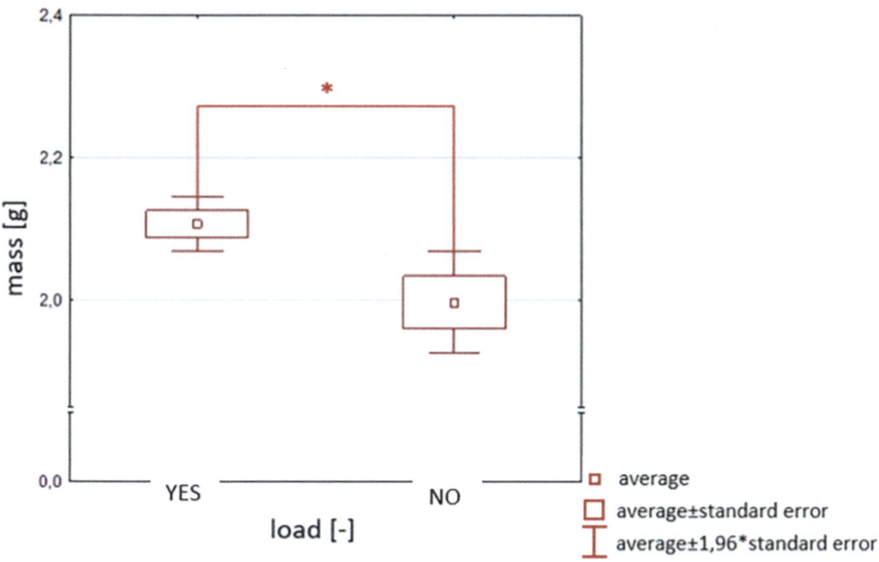

Fig. 4.4 Comparison of the weight of GFRP10_3M_O and GFRP10_3M_BO after 24 h of water absorption: Yes—samples incubated with load. No—samples incubated without unloaded; *—statistically significant differences ($p < 0.05$)

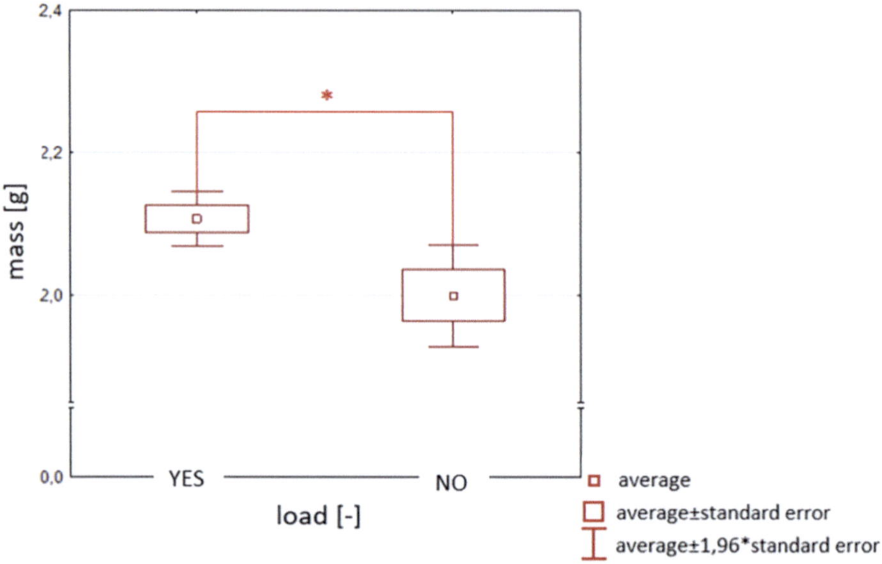

Fig. 4.5 Comparison of the weight of GFRP10_3M_O and GFRP10_3M_BO after 48 h of water absorption: Yes—samples incubated with load. No—samples incubated without unloaded; *—statistically significant differences ($p < 0.05$)

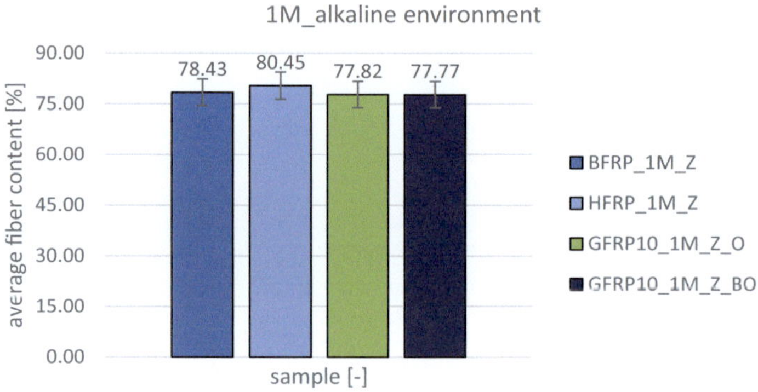

Fig. 4.6 Reinforcement content FRP *rods* after 30 days immersion in alkaline solution

However for basalt rods that were immersed in saline solution, statistical analyses showed the presence of statistically significant differences in the values of both flexural strength and Young's modulus (Fig. 4.12).

In both the first and second case, a correlation was observed between the results obtained and the data obtained from previous tests: determining porosity and measuring the weight of the FRP rods after immersion in aggressive environments. In comparison to primary weight before immersion, FRP rods incubated in

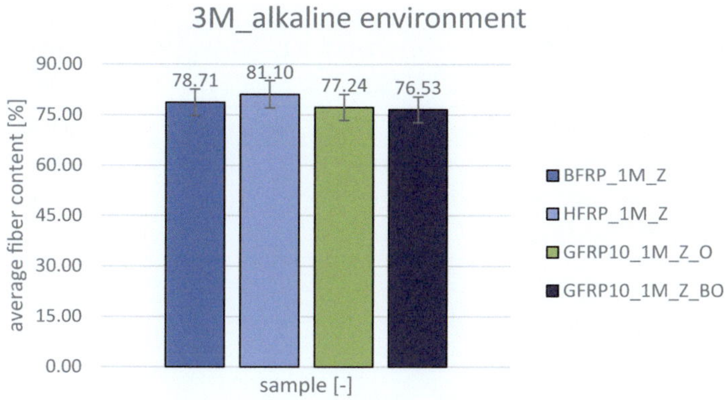

Fig. 4.7 Reinforcement content in FRP rods after 90 days immersion in alkaline solution

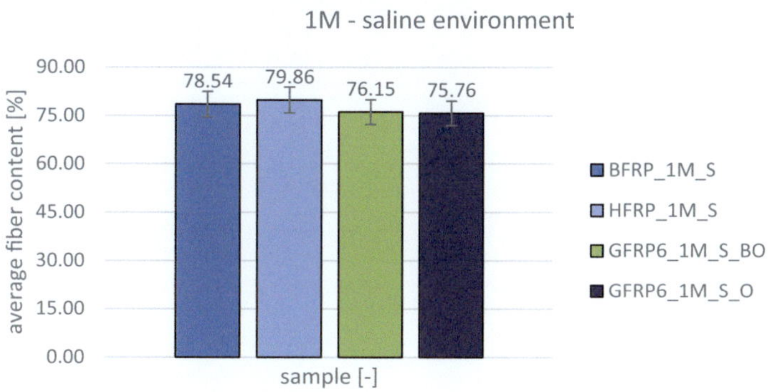

Fig. 4.8 Reinforcement content in FRP rods after 30 days immersion in saline solution

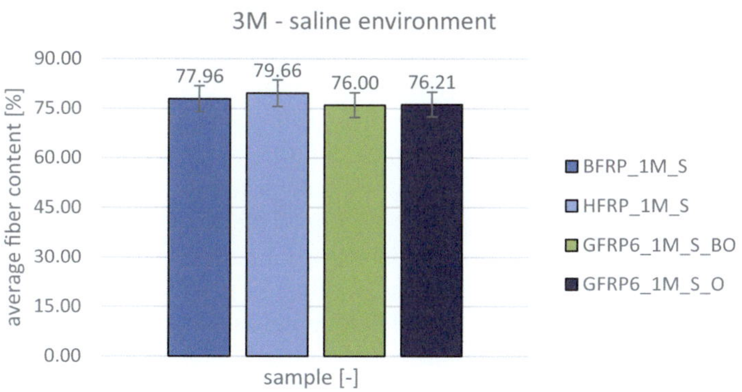

Fig. 4.9 Reinforcement content in FRP rods after 90 days immersion in saline solution

Table 4.3 BFRP four-point bending test results: F—maximum force at which specimens fail [N]; σ—average bending stress [MPa]; ε—average strain [–]; E—average Young modulus [MPa]

Specimen	F [N]	σ [MPa]	ε [–]	E [MPa]
BFRP_1M_S	593.32	503.97	0.059	8969
BFRP_3M_S	468.71	385.09	0.53	6648
BFRP_1M_Z	671.96	490.99	0.06	8448
BFRP_3M_Z	528.60	479.73	0.058	9030

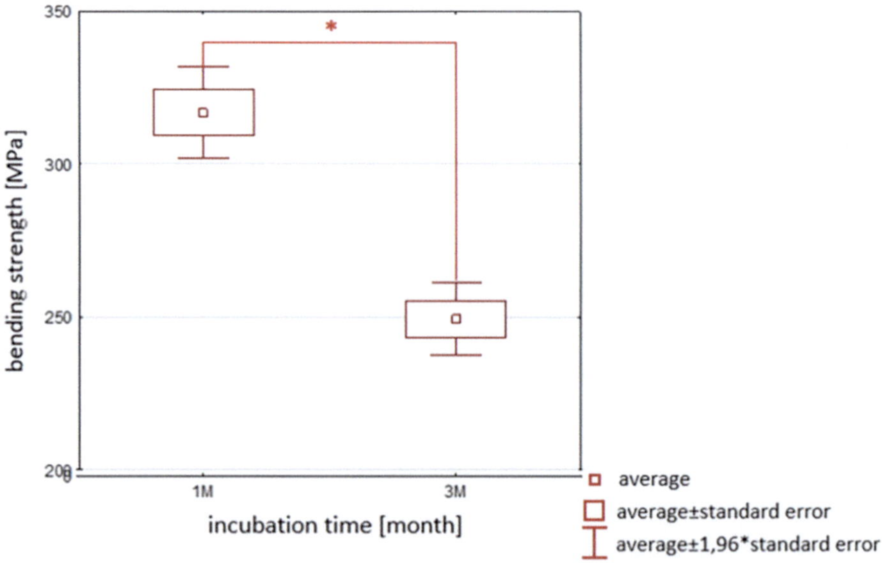

Fig. 4.10 Comparison of flexural strength of BFRP rods incubated for 30 and 90 days in alkaline solution; *—statistically significant differences ($p < 0.05$)

alkaline and saline solution showed an increase in weight after immersion. It was also observed that there were statistically significant differences for the samples tested in the case of open porosity, close porosity and total porosity.

In the case of the stress–strain (σ–ε) curves of the four-point bending test after 30 days immersion, it can be observed that the specimens in the saline solution failed at strain in the range of 0.05–0.06. Similar situation exists for the specimens after 90 days immersion. This results showed that immersion time wasn't important. The results for BFRP rods immersed in the alkaline solution were similar. However, the type of solution affects the strength properties of FRP rods (Fig. 4.13).

Glass FRP rods (GFRP6) also showed a decrease in flexural strength with longer immersion in an aggressive solution. The Young modulus value increased after 30 days of incubated for GFRP rods that were immersed without loads. For samples with load, Young modulus value was decreased (Table 4.4).

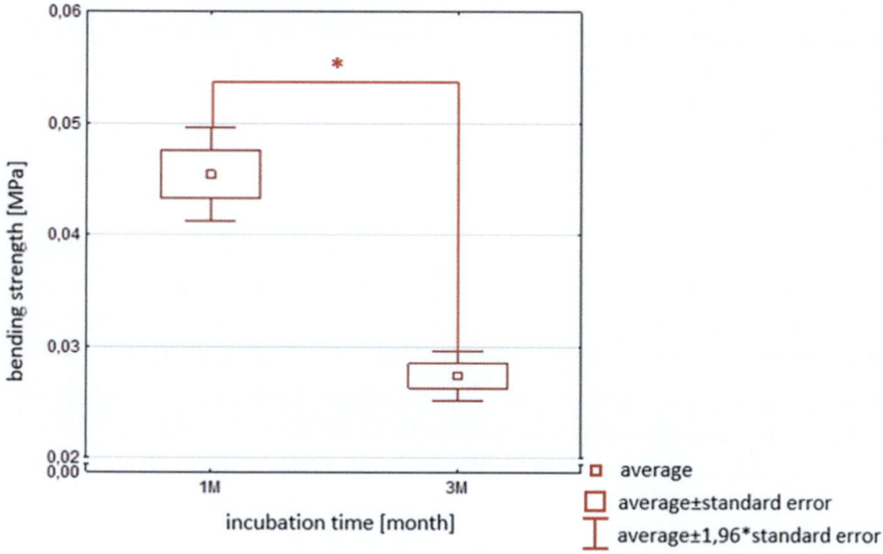

Fig. 4.11 Comparison of strain of BFRP rods incubated for 30 and 90 days in saline solution; *—statistically significant differences ($p < 0.05$)

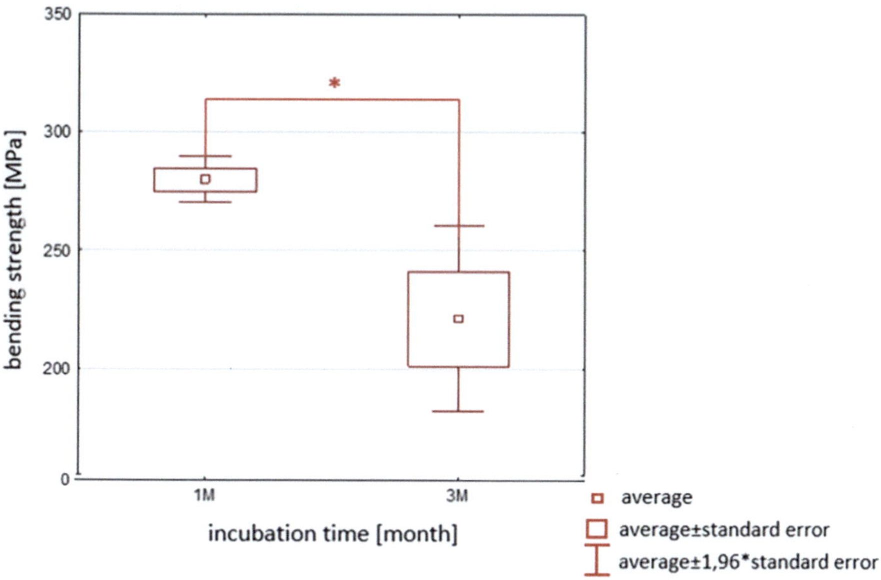

Fig. 4.12 Comparison of flexural strength of BFRP rods incubated for 30 and 90 days in saline solution; *—statistically significant differences ($p < 0.05$)

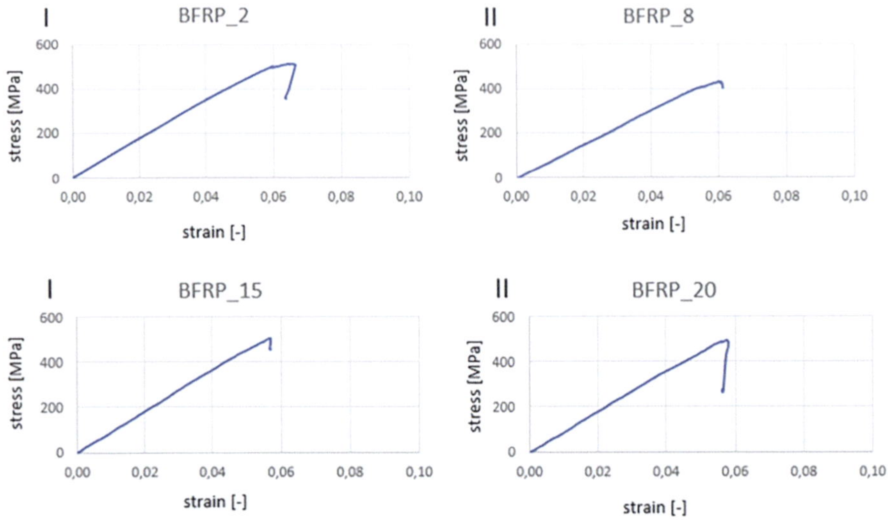

Fig. 4.13 Stress–strain (σ–ε) curve for BFRP specimens after immersion in alkaline solution (BFRP_2 and BFRP_8) and saline solution (BFRP_15 and BFRP_20)

Table 4.4 GFRP6 four-point bending test results: F—maximum force at which specimens fail [N]; σ—average bending stress [MPa]; ε—average strain [–]; E—average Young modulus [MPa]

Specimen	F [N]	σ [MPa]	ε [–]	E [MPa]
GFRP6_1M_S_BO	617.85	483.24	0.07	7582
GFRP6_3M_S_BO	551.61	340.39	0.049	8317
GFRP6_1M_S_O	704.79	469.21	0.062	7497
GFRP6_3M_S_O	567.95	459.52	0.066	7400

For glass rods (GFRP6), statistically significant differences were observed in the values of strain and flexural strength (Figs. 4.14 and 4.15). As in the case of basalt rods, the results obtained from the strength test, the determination of porosity, and the change in the weight of the rods after incubation were also correlated for glass rods (GFRP6).

Graphs showing the σ–ε dependence of the bending test for specimens incubated without an applied load are shown in Fig. 4.16. It can be seen from them that GFRP6 specimens failed at a strain in the range of 0.06–0.09. These values are similar to those obtained by FRP rods of the same diameter reinforced with basalt fiber. For the specimens that were incubated under load, the curves showing the σ–ε relationship of the bending test show that the specimens failed at strains in the range of 0.05–0.08. In addition, the statistical analyses performed for the two groups showed that there are statistically significant differences between the results obtained, both for bending strength and strain.

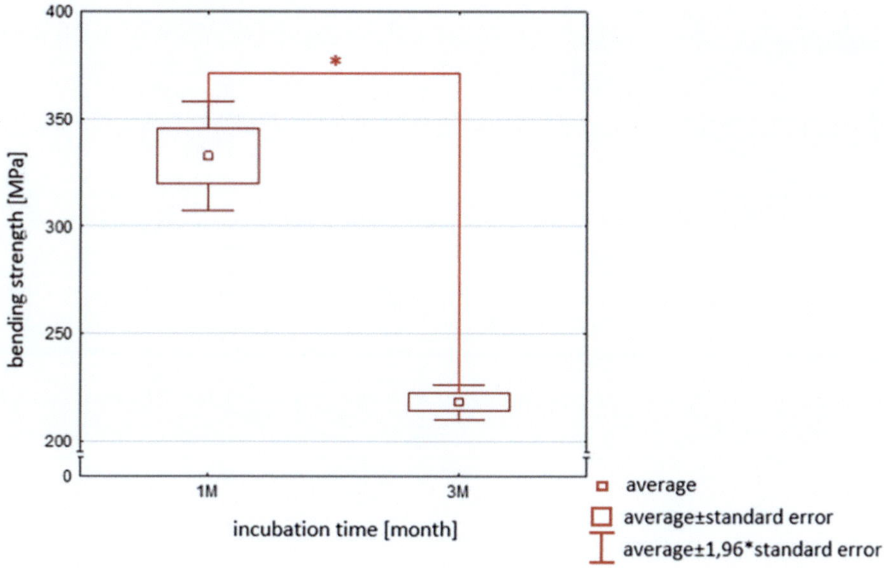

Fig. 4.14 Comparison of flexural strength of GFRP6 rods incubated with load for 30 and 90 days in saline solution; *—statistically significant differences ($p < 0.05$)

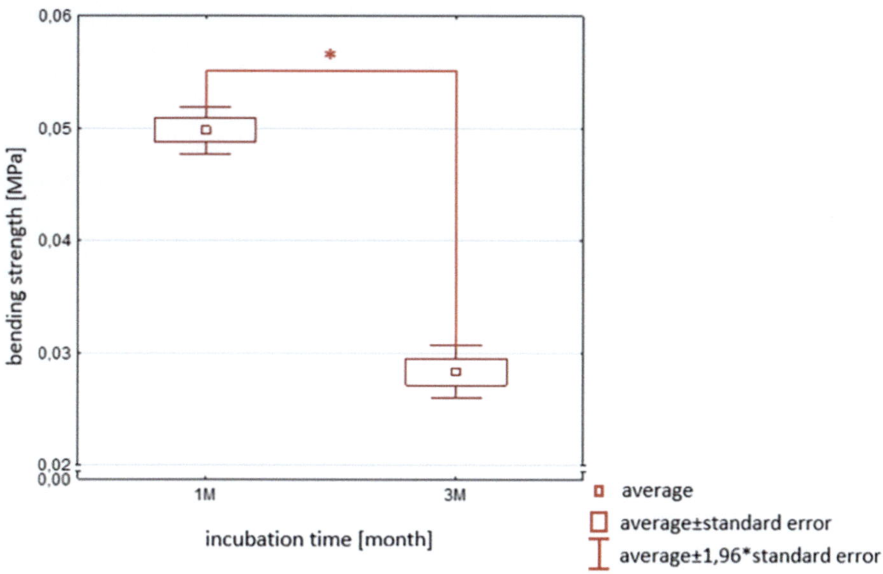

Fig. 4.15 Comparison of strain of GFRP6 rods incubated with load for 30 and 90 days in saline solution; *—statistically significant differences ($p < 0.05$)

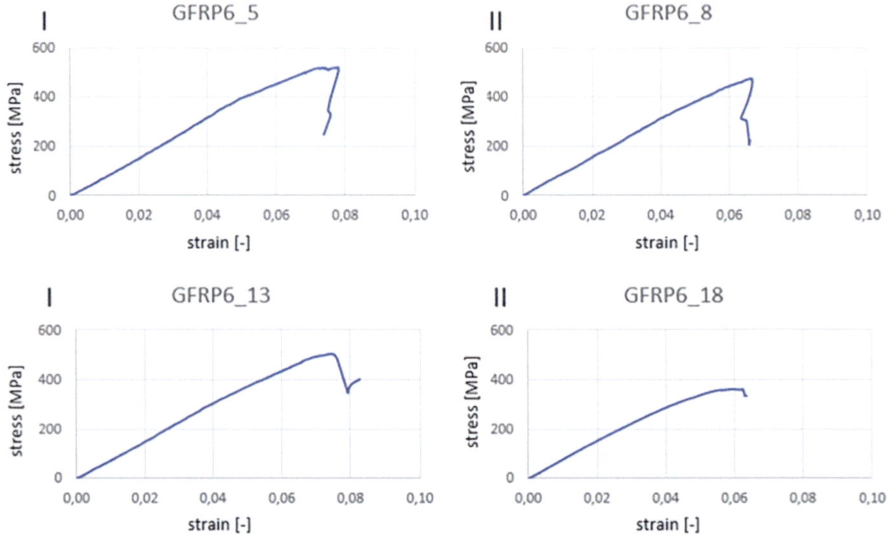

Fig. 4.16 Stress–strain (σ–ε) curve for GFRP6 specimens after immersion in saline solution without load (GFRP6_5 and GFRP6_8) and with load (GFRP6_13 and GFRP6_18)

Glass FRP rods that were immersed in an alkaline environment showed an increase in flexural strength for both loaded (GFRP10_O) and unloaded (GFRP10_BO) rods. Also, the strain value increases with longer immersion time. The situation is different in the case of Young's modulus, for specimens immersed under load increased slightly with longer incubation time. However, for bars immersed without load, the value of Young's modulus decreases significantly after the first month of incubation (Table 4.5).

However, despite the varying test results, the analyses performed did not show the presence of statistically significant differences. Based on these results, it can be concluded that the applied load did not significantly affect the strength parameters of the composite rebars.

The σ–ε curves of the bending test for specimens incubated for 1 month with load clearly indicate that the bars failed at a strain equal to 0.05. For specimens that were immersed for a longer time, failure occurred at a strain of 0.06. In contrast,

Table 4.5 GFRP10 four-point bending test results: F—maximum force at which specimens fail [N]; σ—average bending stress [MPa]; ε—average strain [–]; E—average Young modulus [MPa]

Specimen	F [N]	σ [MPa]	ε [–]	E [MPa]
GFRP10_1M_Z_O	2114.35	421.05	0.054	7948
GFRP10_3M_Z_O	2502.34	442.91	0.057	7998
GFRP10_1M_Z_BO	2481.26	419.85	0.053	8050
GFRP10_3M_Z_BO	2623.95	448.30	0.057	7450

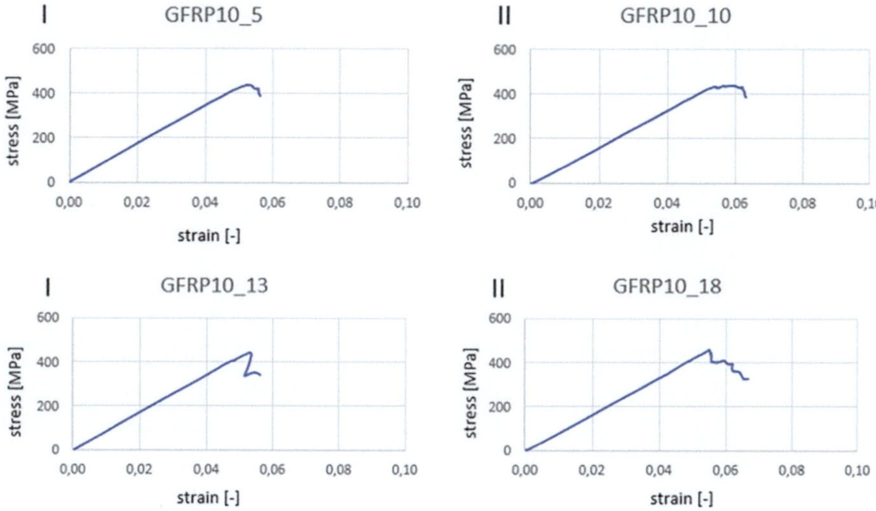

Fig. 4.17 Stress–strain (σ–ε) curve for GFRP6 specimens after immersion in alkaline solution with load (GFRP10_5 and GFRP10_10) and without load (GFRP10_13 and GFRP10_18)

specimens that were immersed without load failed at a strain above 0.06, regardless of the length of incubation (Fig. 4.17).

Hybrid rods (HFRP) that were immersed in a saline environment were characterized by an increase in flexural strength and Young's modulus values with longer incubation time. A similar relationship was observed for bars immersed in an alkaline environment (Table 4.6).

The values of the average strain at which the bars failed decreased with longer immersion time. Analyzing the curves (Fig. 4.18), it can be observed that both hybrid rods immersed in saline and alkaline environments failed at a strain of 0.04–0.06. There were also no differences that could be due to the length of incubation of the samples.

The results obtained clearly agree with what can be found in literature reports. When samples were incubated in aggressive environments, mostly alkaline and saline, groups of researchers observed a decrease in flexural strength. During longer

Table 4.6 HFRP four-point bending test results: F—maximum force at which specimens fail [N]; σ—average bending stress [MPa]; ε—average strain [–]; E—average Young modulus [MPa]

Specimen	F [N]	σ [MPa]	ε [–]	E [MPa]
HFRP_1M_S	1192.73	428.92	0.06	8067
HFRP_3M_S	1284.67	471.28	0.05	8329
HFRP_1M_Z	1388.14	466.12	0.05	7984
HFRP_3M_Z	1308.95	471.29	0.06	8499

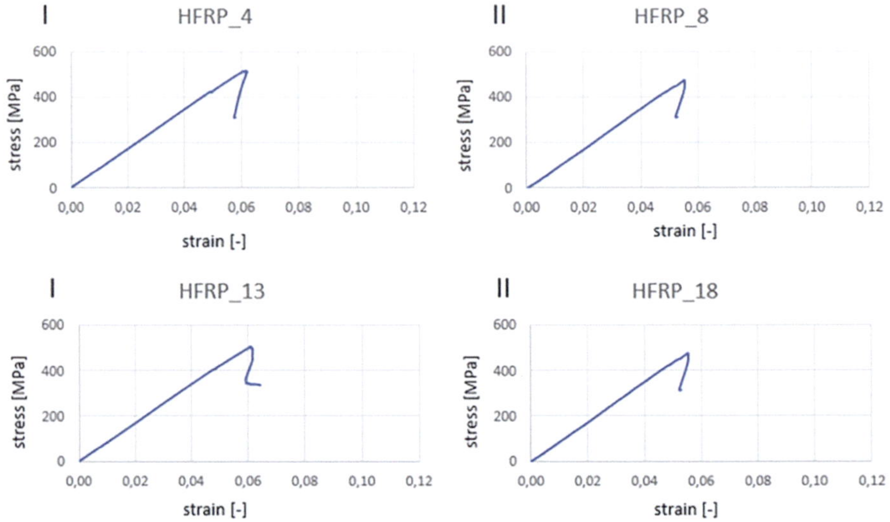

Fig. 4.18 Stress–strain (σ–ε) curve for HFRP specimens after immersion in alkaline solution (HFRP_4 and HFRP_8) and saline solution (HFRP_13 and HFRP_18)

incubation periods it has been observed for BFRP rods that there is a rapid decrease in flexural strength up to a certain point after which it becomes slower.

Based on the results obtained from the four-point bending test, Young's modulus of the specimens was determined. From the data presented, it can be seen that in the case of basalt bars with increasing incubation time, the value of Young's modulus decreased in particular for bars immersed in a salty environment. A similar state of affairs has been observed by many researchers [1–3]. However, the researchers also noted that for rods that were incubated in an environment at a temperature lower than 50 °C the value of Young's modulus did not decrease, but increased.

However, as for basalt rods incubated in an alkaline environment this, study shows a slight increase in Young's modulus value after 3 months of immersion. In the case of glass rods incubated in a saline environment, regardless of whether the samples were immersed with or without load an increase in Young's modulus values was observed after 3 months. The results obtained agree with only part of the literature reports. In a study [4, 5] also observed that during incubation up to 100 days, there is an increase in Young's modulus while with longer immersion time the value begins to decrease. In other articles, the studies presented show a completely different trend—the value of Young's modulus begins to decrease from the very beginning of incubation while in many cases it stabilizes or slightly increases around day 200 [4, 6, 7].

The situation is reversed for glass rods incubated in an alkaline environment—here the fact of whether the samples were immersed with or without loading makes a difference. In the case of incubation without load, it was observed that the value of Young's modulus decreased after 3 months. On the other hand, for bars that were immersed under load, no significant changes were observed in the value of Young's modulus. The results obtained are in line with what can be found in the literature but while in the case of the study by Wang et al. [8] the decrease was small up to 100 days of incubation, in the case of the study by Abbasi [9] the reduction in Young's modulus value was already significant. However, there are also studies that show that there is a decrease in Young's modulus value in the first 200 days while later it stabilizes and even increases slightly [4].

4.4 Conclusions

Samples incubated for three months in aggressive environments tended to show an increasing trend regardless of what the incubation conditions were (with or without load). Only in the case of GFRP10 rods incubated under load was a slight decrease in the weight of the samples observed after 48 h of water absorption.

In the four-point bending test, basalt (BFRP) specimens showed a decrease in flexural strength after prolonged incubation, regardless of the environment in which the rods were immersed. For the other groups, the decreases in strength were insignificant after three months and sometimes the flexural strength increased. It can be clear-cut concluded that basalt rods degrade very quickly with a consequent deterioration in strength properties as well. In the case of hybrid rods (HFRP), it was the time that significantly affected the results obtained, not the environment in which the rods were immersed (increasing flexural strength and Young's modulus values with longer incubation time). On the other hand, for fiberglass-reinforced bars (GFRP6 and GFRP10), which were incubated both with and without loading, no major changes in flexural strength were observed. Time was also of the essence in these groups as slight decreases in strength were observed after three months. Immersion with or without loading no longer had such a large effect. One should consider whether a longer incubation time would be a better solution in this case.

Changes in pH indicate that the fibers themselves are corroding. From the results presented, it can be concluded that a short contact of the rods with an alkaline environment does not cause significant damage, while a longer one does. In the case of a salty environment, it can be observed that the short time for which it acts causes greater damage to the rods, but prolonged contact is unlikely to cause significant damage.

Acknowledgements The research was financed by the National Center for Research and Development of Poland (NCBiR) grant number LIDER/40/0219/L201010/18/NCBR/2019.

References

1. Wang Z, Zhao XL, Xian G, Wu G, Singh Raman RK, Al-Saadi S, Haque A (2017) Long-term durability of basalt- and glass-fibre reinforced polymer (BFRP/GFRP) bars in seawater and sea sand concrete environment. Constr Build Mater 139(15):467–489
2. Wang X, Wu G, Wu Z, Dong Z, Xie Q (2014) Evaluation of prestressed basalt fiber and hybrid fiber reinforced polymer tendons under marine environment. Mater Des 64:721–728
3. Wu G, Wang X, Wu Z, Dong Z, Xie Q (2015) Degradation of basalt FRP bars in alkaline environment. Sci Eng Compos Mater 22(6):649–657
4. Al-Salloum YA, El-Gamal S, Almusallam TH, Alsayed SH, Aqel H (2013) Effect of harsh environmental conditions on the tensile properties of GFRP bars. Compos B Eng 45(1):835–844
5. Wang Z, Zhao X-L, Xian G, Wu G, Singh Raman RK, Al-Saadi S (2018) Effect of sustained load and seawater and sea sand concrete environment on durability of basalt- and glass-fibre reinforced polymer (B/GFRP) bars. Corros Sci 138:200–218
6. Heshmati M, Haghani R, El-Emrani M (2017) Durability of bonded FRP—to—steel joints: effects of moisture deicing salt solution temperature and FRP type. Compos B Eng 119:153–167
7. Mufti AA, Banthia N, Benmokrane B, Boulfiza M, Newhook JP (2007) Durability of GFRP composite rods. Concr Int 29(02):37–42
8. Wang W, Meng Y (2009) Durability of FRP bars exposed to alkaline corrosion environment. China Synthetic Resin Plast 26:27–30
9. Abbasi A, Hogg PJ (2005) Temperature and environmental effects on glass fibre rebar: modulus strength and interfacial bond strength with concrete. Compos B Eng 36(5):394–404

Chapter 5
Bond Behaviour of FRP Rebars Embedded in Concrete

Cristiane Lopes, Bruno Pedrosa, Paweł Zielonka, Anna Wybraniec, Tomasz Socha, Arkadiusz Denisiewicz, Krzysztof Kula, José A. F. O. Correia, and Grzegorz Lesiuk

Abstract The performance of the concrete beam depends on many aspects including reinforcement type, their architecture in beam, bond behavior between concrete and reinforcing bars and many others. However, the aspect in this chapter is placed in bond behavior of the composite bars and concrete. This behavior is crucial in terms of effective performance of concrete beam, weak bond connection results in pre-failure of the member. To investigate the adhesive behavior the pull-out tests were performed for pultruded composite rebars and experimental data were correlated with available models that predict this behavior. A good agreement of experimental results with analytical models was found.

Keywords Mechanical performance · FRP bars · Unidirectional composites · Elastic properties

5.1 Introduction

In this chapter, the bond behavior of FRP rebars embedded in concrete is addressed. This topic has been the focus of many research papers namely because there is no standardization of the surface treatment for these rebars. The interaction of the two

---

C. Lopes (✉)
Department of Structural Engineering, Federal University of Minas Gerais, Belo Horizonte, Brazil
e-mail: cristianecarolinelopes@yahoo.com.br; hermes@dees.ufmg.br

B. Pedrosa · J. A. F. O. Correia
CONSTRUCT & INEGI/LAETA, Faculdade de Engenharia, Universidade do Porto, Porto, Portugal

P. Zielonka · A. Wybraniec · G. Lesiuk
Faculty of Mechanical Engineering, Wroclaw University of Science and Technology, Wroclaw, Poland

T. Socha · A. Denisiewicz · K. Kula
Institute of Building, University of Zielona Góra, Zielona Góra, Poland

G. Lesiuk et al. (eds.), *Composite Rebars for Reinforced Concrete Structures*, Structural Integrity 34, https://doi.org/10.1007/978-3-031-96292-9_5

materials is a decisive factor in the overall behavior of the structures because the effectiveness of FRP rebars in concrete structures depends on the bond strength (axial pull-out) and flexural strength [1]. The interaction between FRP rebars to concrete is very different from steel rebars since this parameter is mainly dependent on friction forces developed in the surface. The bond between the FRP rebar to concrete is influenced by many parameters, such as: rebar diameter, fiber type, concrete cover, concrete strength, and many others. An experimental campaign was performed aiming to characterize the bond behavior of developed hybrid FRP rebars embedded in concrete. Results were then analyzed considering an analytical approach.

The typical bond behavior of FRP rebars embedded in concrete is described by the relation between the bond stress and the slip displacement as presented in Fig. 5.1. This bond-slip curve can be divided in 4 stages. The first one is characterized by an increasing bond stress with zero slip displacement. This behavior is controlled by the chemical adhesion effect between the materials. When the chemical adhesion is exceeded, the rebar starts to slip and it is the beginning of stage 2. The bond behavior in this stage is linear and it is determined by the friction effect which is the increase of bond stress in the first stages of movement caused by the shear stresses that appears due to contact between the rebar and concrete surface. Then, the non-linear behavior between the end of stage 2 and the maximum bond stress is labelled as stage 3. It is ruled by mechanical interlocking effect. This phase is influenced by the localized forces caused by the relative displacement due to the surface deformation of the rebar. And finally, after the maximum bond stress is reached, a constant branch is found with increasing slip deformation with no variation of bond stress. This final stage is controlled by the residual stress effect.

The bond behavior between FRP rebars and concrete is different from the bond behavior of steel rebars to concrete in many aspects. First of all, the bond strength of FRP rebars is usually within 40–100% of the bond strength of steel rebars [3]. Also, the bond failure of steel rebars typically occur in the concrete surrounding

Fig. 5.1 Bond stress slip schematic diagram [2]

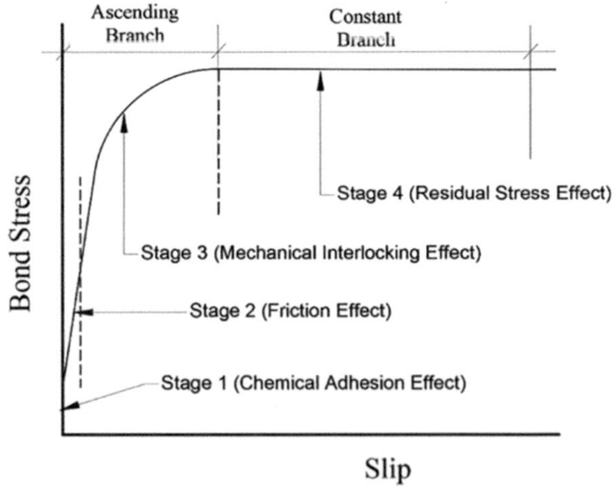

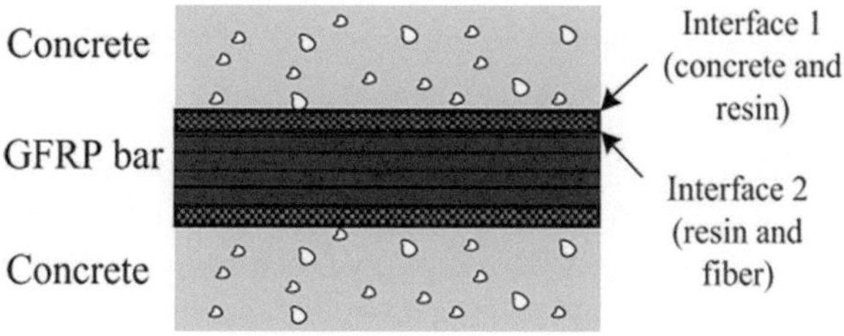

Fig. 5.2 Bond failure interfaces for FRP rebar and concrete [5]

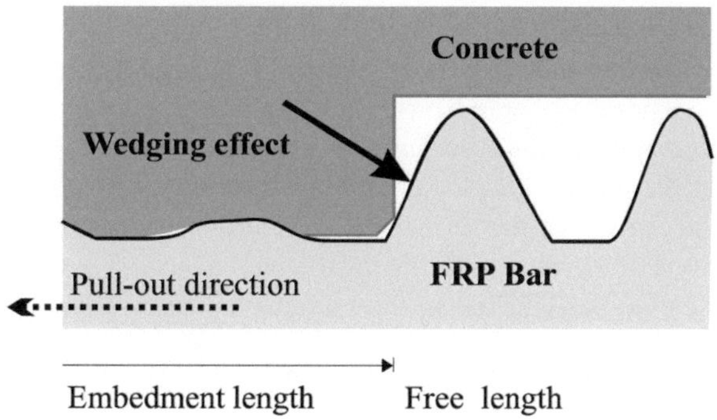

Fig. 5.3 Detail on the wedging effect on FRP rebars [4]

the rebar[4] while the bond failure of FRP rebars can occur not only in the interface between concrete and resin (Interface 1 in Fig. 5.2) but also in the resin fibers interface (Interface 2 in Fig. 5.2) [5].

Another important aspect is that FRP rebars usually present higher values of residual bond strength due to the contribution of the rebar undamaged section which is pulled into the embedment length in the concrete block. However, this residual bond strength is not a realistic resistance since it is only the result of the wedging action of the undamaged part of the rebar as illustrated in Fig. 5.3. On the other hand, steel rebars are not affected by the entering of the undamaged part of the rebar into the concrete because the bond failure typically occurs on the surface of the concrete and, when the rebar enters the block, the concrete surface is already highly damaged and does not offer resistance to sliding [4].

5.2 Governing Parameters

5.2.1 Rebar Surface Treatment

The bond behavior between FRP rebars and concrete is highly influenced by the rebar surface treatment. Smooth rebars have only 10–20% of the bond strength when compared with rebars with surface treatment [4]. Due to the lack of standardization related to surface treatments in rebars, the names used in the literature to designate each treatment also vary significantly, for example: helically wrapped, indented, and ribbed [6].

In the literature there are different ways of generating deformations on the surfaces of FRP rebars. In Achillides and Pilakoutas[4] the rebars used in the pullout tests had their surface deformed through a peel-ply process added to the surface during the pultrusion process. This peel ply process was described by Mcsweeney and Kropa [7] as an extra layer of a synthetic material applied to the outer surface of the composite during manufacture. This material absorbs part of the resin from the surface of the composite and becomes part of it during the curing process. The material used in the peel ply is subsequently removed, causing deformations in the outermost layer of the composite, which became its surface rough for bond purposes.

The sand coated process is also used for increase the bond of the FRP rebars to concrete, and it is made by applying an extra layer of epoxy resin with sand [8] in the surface of the rebar. Helically wound rebars are manufactured by wrapping the smooth rebar in a helical strand that is subsequently removed in the end of the manufacture process living the surface of the rebars deformed. In the helically wrapped surface the smooth rebars is wrapped by a bundle of fibers with epoxy resin and this extra layer of fibers is maintained after the process. The process for making indented rebars is done by cutting into the perimeter of the smooth bar, leaving a undulating surface [9], however, this process causes a reduction on the rebar cross section and the destruction of the fibers on the rebar surface [8] (Fig. 5.4).

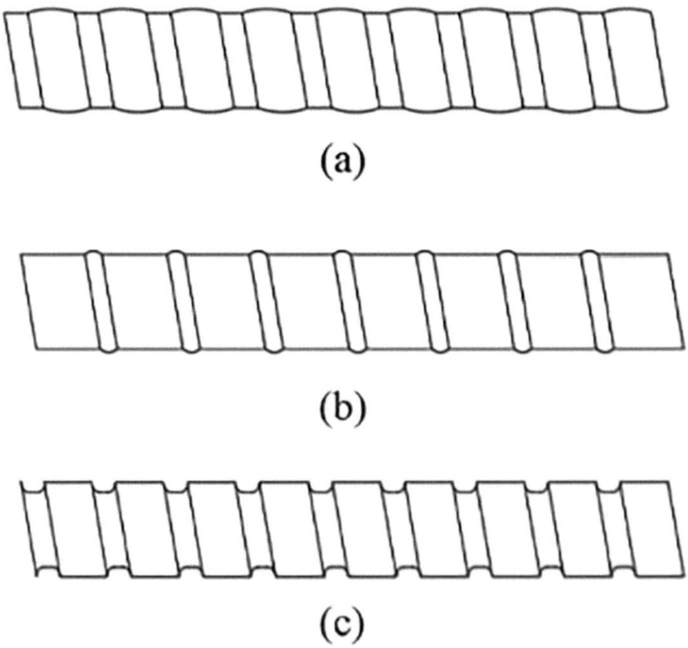

Fig. 5.4 FRP rebars surface treatments. **a** Helically wound, **b** helically wrapped, **c** indented [9]

Different types of treatment helps to improve different bond stages. For example, sanded with spiral wound with fibers improve the adhesion and mechanical interlock since sanded treatment contribute for the chemical adhesion and ribbed treatment increases the mechanical interlock [10].

The surface treatment also changes the after-peak bond behavior of the FRP rebars. In the case of sand coated surface treatment, once the maximum bond strength is achieved, the sanded coated surface detaches from the rebar creating an abrupt drop in the bond stress. This abrupt decay is smoother in cases where normal strength concrete is used. For high strength concrete, the entire sand coated layer gets debonded from the rebar, suddenly [11].

Comparing the effects of the surface treatment used to increase the maximum bond strength, Henin et al. [12] found that using a second layer of sand in BFRP rebars increases the maximum bond strength by 80%. Furthermore, the use of coarse sand coating is more effective than fine sand coating. For fine sand coating, the bond forces between the rebars and concrete is transferred just by friction while for coarse sand coating, bond forces are created by friction and mechanical interlocking. Comparing also the indented, ribbed and helically wrapped surface, Başaran at el. [6] found that an higher performance is achieved with the helically wrapped surface treatment.

5.2.2 FRP Rebar Diameter

The bond between FRP rebar and concrete decreases with the increase of rebar diameter. There are some explanations for this behavior, such as, shear lag, embedment length, gaps in the mixture and also the Poisson effect [6]. The shear lag occurs due to the non-uniform normal stress distribution through the cross section of the rebar, which leads to a differential movement between the core and surface fibers see Fig. 5.5. This means that, with larger rebar diameters, the difference between the stresses in the core and in the surface is higher, which enhances this effect [4]

FRP rebars with larger diameters need longer embedment lengths to develop the same normal bond stress, and larger embedment lengths leads to a reduction on

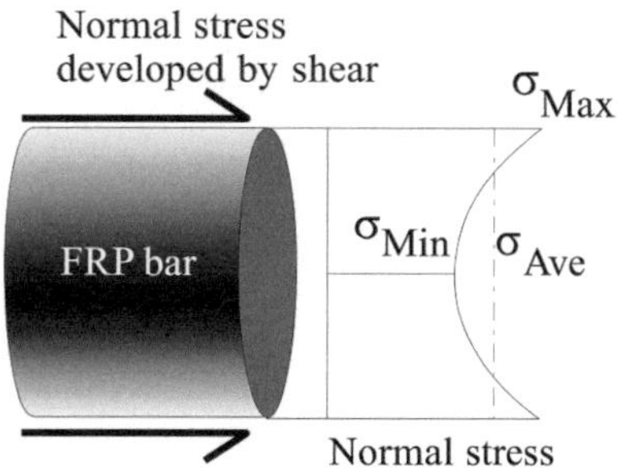

Fig. 5.5 Shear lag effect [4]

the average bond strength [4]. Furthermore, higher diameter can lead to more water accumulated underneath the rebar, which generates a higher number of voids in the mixture and consequently lower bond strength [6]. Finally, in concerns the Poisson effect, the longitudinal stress applied to the rebar in a pullout test leads to a slight reduction on the rebar diameter. This effect is more significative in rebars with higher diameters and it can result in a reduction on friction/mechanical locking stresses [4].

5.2.3 *Fiber Type*

The type of fibers also has an expressive influence on the bond behavior between the FRP rebars and concrete, even if the influence of the surface treatment is considered the most influential parameter [6]. Since bond behavior is affected by many factors, it is difficult to access the effect of the fiber type in the overall bond behavior of the rebar. However, there are some important conclusions in the literature: bond strength of CFRP and GFRP rebars represents around 72% of steel rebars bond strength [4]. When comparing different material types for FRP rebars, carbon-based rebars showed a bond strength 0.9 to 2.8 times the bond strength of glass-based rebars. Results also indicate that basalt-based rebars have around 1.3 to 1.9 times the bond strength of glass-based rebars. When compared to basalt-based rebars, the bond strength of carbon-based rebars is 0.7 to 1.5 times, and when compared to aramid-based rebars, bond strength is between 1.2 and 1.7 times higher [6].

Furthermore, the type of fibers influences the stiffness of the bond-slip curves. GFRP tends to present higher slip values due to its lower elastic modulus [11].

5.2.4 *Concrete Strength*

The load transfer between the rebar and concrete depends on three bond mechanisms: adhesion, bearing and friction. The bond failure can occur in the rebar or the concrete surface. When the failure happens in the concrete surface, it is caused by concrete splitting or shear failure, and in this case, the tensile strength of concrete is a key parameter. The strength of the concrete has a greater influence on the bond behavior of FRP rebar if concrete with a compressive strength below 15 MPa is used, because in this case the bond failure is mainly in the concrete surface and the concrete strength has more impact in the bond strength. For concrete with a compressive strength above 30 MPa, failure normally occurs on the surface of the FRP rebar, so the strength of the concrete does not have a significant impact on the bond strength value [4].

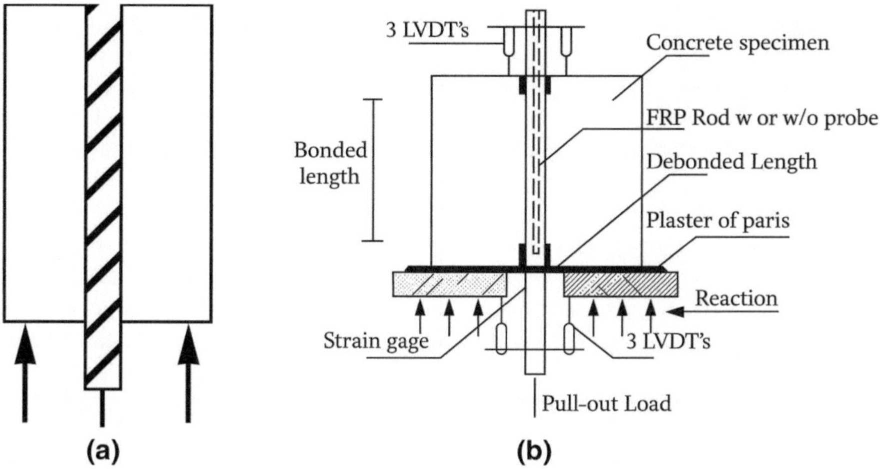

Fig. 5.6 Bond strength determination by pullout testing of FRP rebars [15]

5.3 Experimental Approach

5.3.1 Common Procedures

Various types of test methods are available in literature to assess the bond strength of FRP reinforcement in concrete structures [13]. It's possible to list among them the direct pullout test, the rod-rod pullout test, and the beam pullout test. The direct pullout test, illustrated in Fig. 5.6, consists of measuring the pullout strength of an FRP rebar embedded in a concrete block through a predetermined bonded length. A bond breaker is applied along the first five diameters on the pulling side of the rebar, to minimize any undesired confinement effect that may affect the bond characteristics. The number of test specimens should not be less than five [14].

5.3.2 Experimental Tests

Pull-out tests are a simple and economical test procedure to evaluate the bond performance of a rebar embedded in concrete. The experimental program described in this chapter was based on the recommendations of ACI 440.3R [15]. The bond behavior of hybrid FRP rebars were analyzed designated as Wroclaw 2—see Fig. 5.7. Wroclaw 2 is made of glass (95%) and carbon (5%) fibers with a fiber content of 55%. The rebars had a helically wrapped surface treatment and the overwrapping material used was carbon-based fibers. Geometrical and mechanical properties of the rebar is presented in Table 5.1. The bond-slip relationship was assessed by means of 3 pull-out tests for each rebar type. Furthermore, three samples were tested to define the compression strength of the concrete. The test results indicate an average cubic strength of 67.81 MPa with a standard deviation of 0.52 MPa. According to Eurocode 2 [16], this concrete is classified as C50/60. The concrete composition was based on CEM II/A-V 42.5 R cement and aggregate with 16 mm of maximum size. This concrete strength is a guarantee that the bond failure occur on the surface of the FRP rebars and not inside the concrete.

Concrete cubes had 150 mm side and the rebars were positioned in its center. The FRP rebars extended 500 mm beyond the concrete block for the proper attachment of the anchoring system placed at the end of the rebar where the force was applied. The geometry of specimens and the identification of rebar positioning into the concrete cube is presented in Fig. 5.8. The embedded length adopted was 5 times the rebar diameter (5 $\varnothing_f$) as recommended in the standard. This is the region where FRP rebar is in contact with concrete.

The casting process was made with a wooden structure. A PVC tube was used to guarantee that the bond length is 5 $\varnothing_f$. Figure 5.9 shows the preparation for concrete casting.

The layout for the pullout test is presented in the Fig. 5.10. A Universal Testing Machine was used to perform the pull-out test and the load was applied at the rebar at a rate of 0.1 mm/min. Three linear variable displacement transducers (LVDT) were used to measure the slip of the rebar relative to the concrete cube: one LVDT was attached in the unloaded end of the rebar (LVDT_1) [the unloaded end is the end of

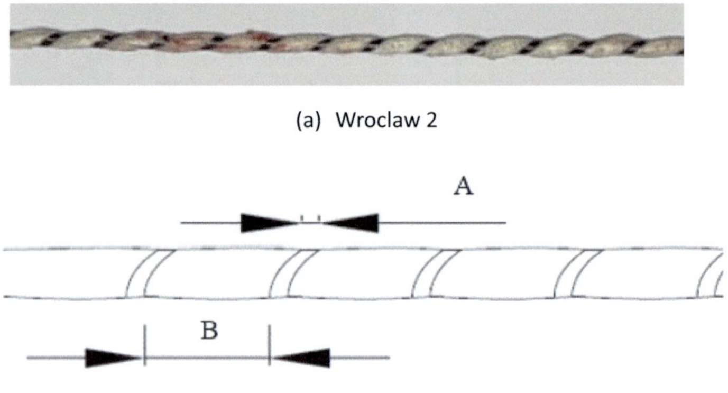

(a) Wroclaw 2

(b) Wroclaw 2

Fig. 5.7 Surface treatment *detail* of FRP rebars Wroclaw 1 and Wroclaw 2

Table 5.1 Characterization of hybrid FRP rebars

Rebar type	Diameter [mm]	Modulus of elasticity [GPa]	Tensile strength [MPa]	Rib width (A) [mm]	Rib spacing (B) [mm]
Wroclaw 2	8.3	32	500	3	21

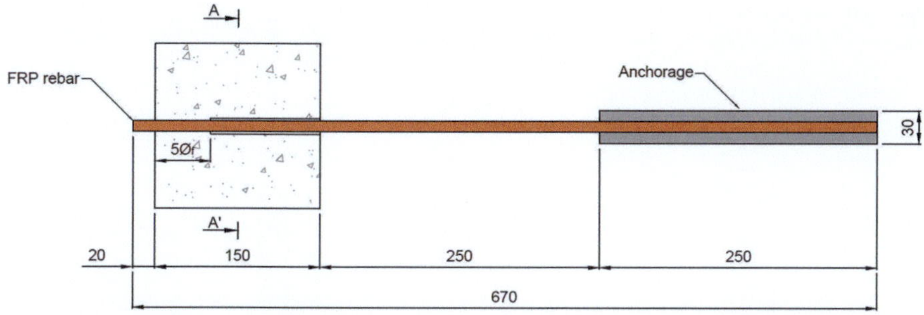

Fig. 5.8 Characterization of pullout specimens

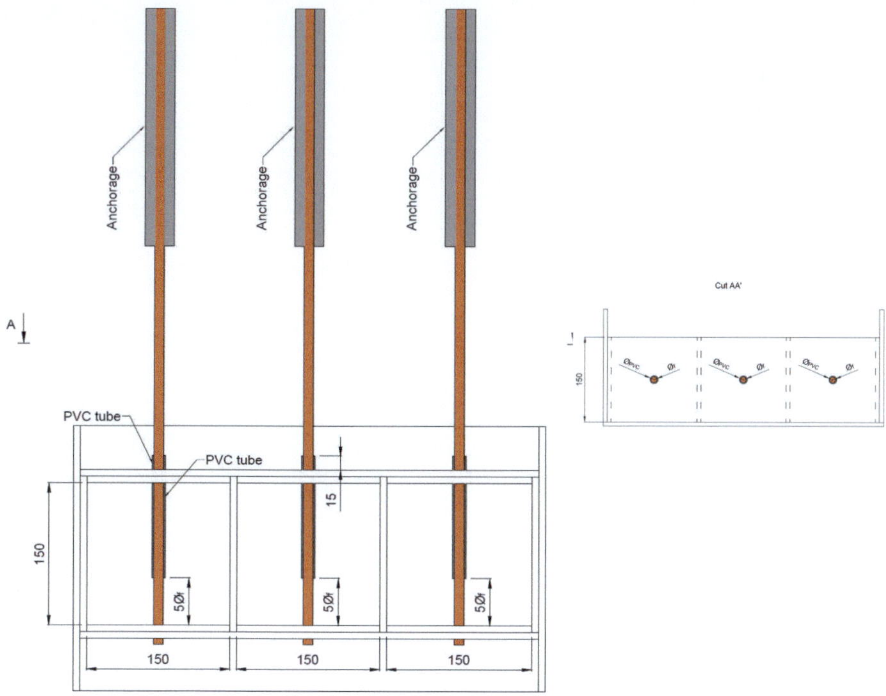

Fig. 5.9 Preparation of specimens—concrete casting

the rebar opposed to the end where the load is applied], one in the bottom surface of the concrete cube (LVDT_2) and one on the loaded end of the rebar (LVDT_3), the loaded end is the end of the rebar were the force is applied.

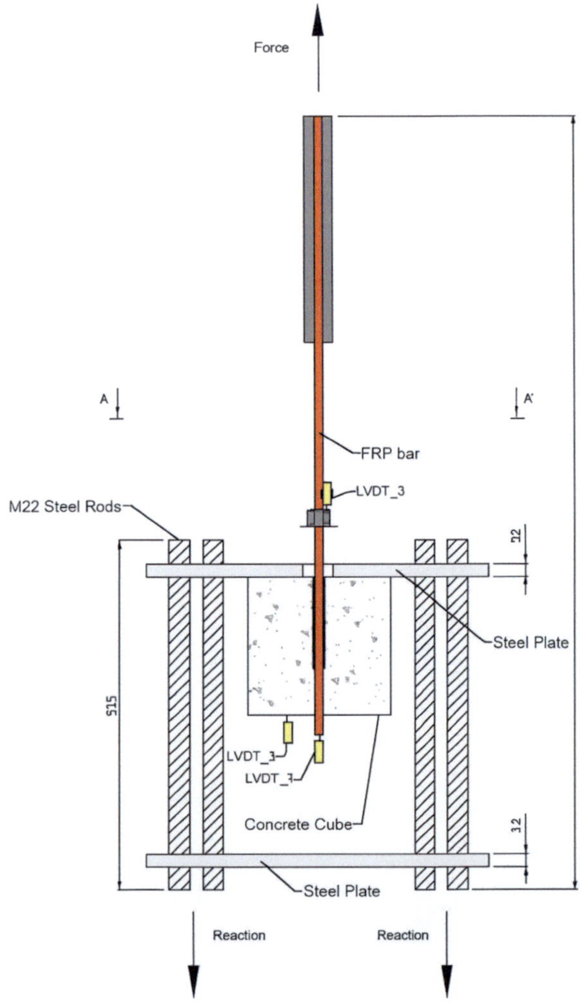

Fig. 5.10 Layout of pullout test

5.4 Experimental Results

The average bond stress, τ, was determined according to Eq. (5.1) where d_b is the nominal diameter of the rebar and L_a is the embedment length. The slip displacement was calculated considering the measurement on the loaded end (LVDT_3) minus the elongation, S_c, Eq. (5.2) of the FRP rebar, considering the distance between LVDT_3 and the surface of the concrete block.

$$\tau = \frac{F}{\pi d_b L_a} \tag{5.1}$$

$$S_c = \frac{FL_c}{E_L A} \tag{5.2}$$

The interaction between concrete and the FRP rebars is characterized by bond stress versus slip curve where initially there is a large increase in bond stress and a small increase in slippage. When the maximum bond stress is achieved, the slippage increase significantly and the bond stress no longer increases.

In Fig. 5.11, a typical load versus slip curve is presented for one specimen tested considering the loaded and unloaded parts of the rebars. The slip of the loaded end of the rebar begins in the early stages of loading. Free end displacement only occurs in the final stages of loading when the bond strength between the rebar and concrete is exceeded. In the results shown below, the unloaded end only begin to slip at around 85% of the maximum bond strength.

We had three samples for rebar type, Wroclaw2-Ø8 (Fig. 5.12). As it is possible to observe, there is good agreement between the results.

Figure 5.13 indicates that bond failure occurred on the rebar surface since a reduction in the diameter of the rebar was observed, as well as the presence of a white residue caused by damage in the resin. These finds suggested that the bond failure happened because of because the shear strength between the resin and the fibers was reached, which was expected due to the use of high-strength concrete. In addition, it can be seen that, although there was a significant damage in the surface of the rebar, the fibers used to form the helically wrapped surface of the rebar was still present and did not detach from the rebar.

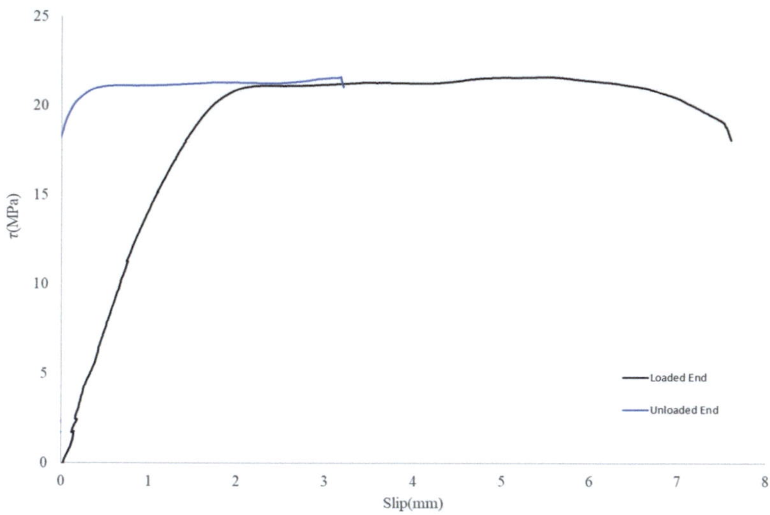

Fig. 5.11 Typical bond stress—*slip* relation for loaded and unloaded rebar end

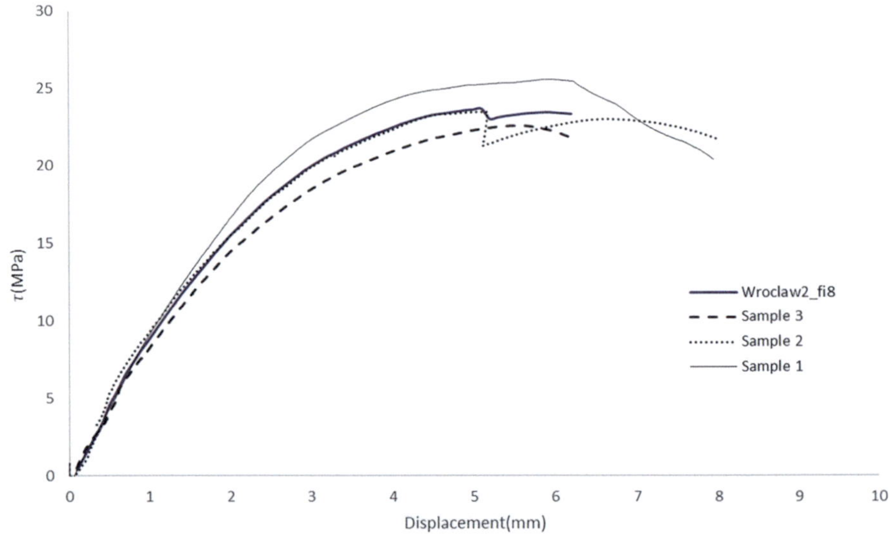

Fig. 5.12 Bond slip graph for Wroclaw2-Ø8 rebars

Fig. 5.13 FRP rebar surface after pullout tests

5.5 Analytical Analysis

There are several analytical models to describe the bond behavior of FRP rebars. In the literature is possible to find many authors [11] comparing experimental bond results with analytical models proposed and adapted to describe the bond behavior of the FRP rebars to concrete. In this section one example of the experimental results obtained in the experimental campaign conducted by the authors is compared with some well-known bond behavior models. The following models were selected and will be briefly presented later: The model proposed by Eligehausen, Popov and Bertero, called BPE model, the Malvar model, the model developed by Consenza, Manfred and Realfonzo referred as CMR model and a modification of the model proposed by Eligehausen, Popov and Bertero named as mBPE.

BPE Model (1982):

Model proposed by Eligehausen, Popov and Bertero for steel rebars. In this model, the ascending branch of the curve is described as presented in Eq. (5.3) from the beginning of the test until S1. The first branch is followed by a constant region with the bond stress equal to the maximum bond strength up to a slip equal to S2. Finally, there is a descending branch with a defined slope (Fig. 5.14).

$$\tau = \tau_1 \left(\frac{s}{s_1} \right)^{\alpha} \text{ For } 0 \le s \le s_1 \tag{5.3}$$

where: τ_1 is the peak bond stress, s_1 is the slip corresponding to the peak bond stress and α is a curve fitting parameter. The parameter α must be less than 1 to be physically meaningful.

Malvar Model (1995):

This is one of the first models developed for the bond-slip behavior of FRP rebars and concrete [18]. The model has two empirical constants depending on the rebar type.

For the rebars used in experimental campaign, empirical constants F and G were calibrated as 1.7 and 1.4, respectively.

$$\tau = \tau_m \frac{F\left(\frac{s}{s_m}\right) + (G-1)\left(\frac{s}{s_m}\right)^2}{1 + (F-2)\left(\frac{s}{s_m}\right) + G\left(\frac{s}{s_m}\right)^2} \tag{5.4}$$

where: τ_m is the peak bond stress, s_m is the slip corresponding to the peak bond stress.

Modified BPE model (1996):

The modified BPE model presents the same ascending branch as the BPE model but without the constant part of the curve. The ascending branch is followed by a descending linear branch with the slop $p\left(\frac{\tau_1}{s_1}\right)$ (Fig. 5.15).

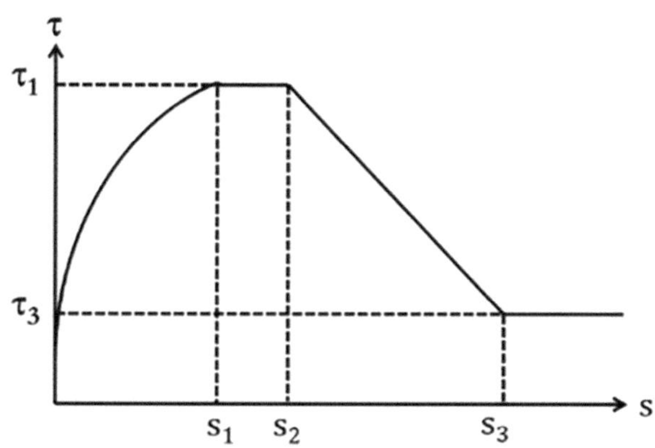

Fig. 5.14 BPE model curve [17]

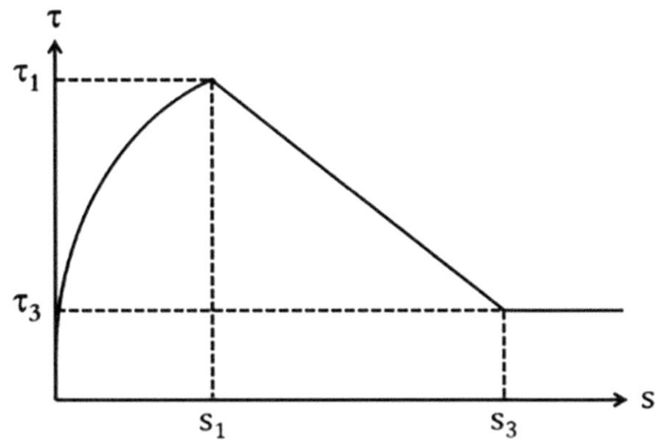

Fig. 5.15 Modified BPE model curve example [17]

$$\tau = \tau_1(1 - p\left(\frac{s}{s_1} - 1\right), s > s_{\max} \tag{5.5}$$

where: τ_1 is the peak bond stress, s_1 is the slip corresponding to the peak bond stress and p is the parameter that account for the softening.

CMR model (1997):

The model developed by Consenza, Manfred and Realfonzo is a modification of the ascending branch of the bond-slip curve proposed in the BPE model.

$$\tau = \tau_m\left(1 - \exp^{\left(\frac{s}{s_r}\right)}\right)^{\beta} \tag{5.6}$$

where: τ_m is the peak bond stress, and s_r and β are curve fitting parameters obtained from the experimental data as -1.1 and 2, respectively. Comparing experimental and analytical results (see Fig. 5.16), it is possible to observe a good agreement, especially for the Malvar and CMR models, which were able to represent the average values of all the samples analyzed. The BPE and Modified BPE models were not able to satisfactorily represent the ascending section of the bond behavior curves of these rebars.

Below is an example of the results obtained for the pullout tests carried out in the experimental campaign conducted by the authors compared to the analytical models described above. As can be seen, the results obtained by the CMR and Malvar curves were closer to the results obtained experimentally and can therefore be considered more efficient in representing the behavior of these rebars.

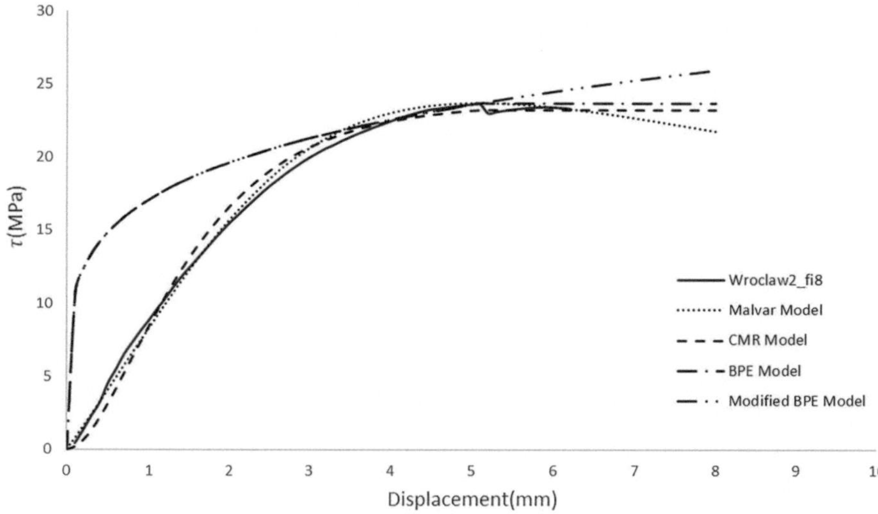

Fig. 5.16 Comparison between analytical and experimental results for Wroclaw2-fi8 rebar

5.6 Conclusions

The bond behavior of FRP rebars to concrete is an important and recurrent topic on FRP rebars, either because of the lack of standardization for the surface treatment of these rebars, or because of the large number of parameters involved in their behavior (concrete strength, fiber type, concrete cover). The bond behavior of these type of rebars to concrete is usually assessed through pullout tests that are a simple and economical way to analyses the bond behavior of FRP rebars embedded in concrete.

As can be seen from the results presented in this chapter, there is a good agreement for the three samples tested for the rebar Wroclaw 2. Also the behavior of these rebars could be efficiently represented by two of the analytical models analyzed.

Acknowledgements The research was financed by the National Center for Research and Development of Poland (NCBiR) grant number LIDER/40/0219/L-10/18/NCBR/2019.

References

1. Fergani H, Guadagnini M, Lynsdale C, Mias C, Di Benedetti M (2016) Long term performance of GFRP bars under the combined effects of sustained load and severe environments
2. Henin E, Morcous G (2021) Bond behavior of helically wrapped sand coated deformed glass fiber-reinforced polymer (GFRP) bars in concrete. Constr Build Mater. https://doi.org/10.1016/j.conbuildmat.2021.123120
3. Okelo R, Yuan RL (2005) Bond strength of fiber reinforced polymer rebars in normal strength concrete. J Compos Constr 9:203–213. https://doi.org/10.1016/j.conbuildmat.2008.04.011

4. Achillides Z, Pilakoutas K (2004) Bond behavior of FRP bars under direct pullout conditions. J Compos Constr 8:173–181
5. Lee JY, Kim TY, Kim TJ, Yi CK, Park JS, You YC et al (2008) Interfacial bond strength of glass fiber reinforced polymer bars in high-strength concrete. Compos Part B Eng 39:258–270. https://doi.org/10.1016/j.compositesb.2007.03.008
6. Başaran B, Kalkan İ, Beycioğlu A, Kasprzyk I (2022) A review on the physical parameters affecting the bond behavior of FRP bars embedded in concrete. Polymers (Basel) 14. https://doi.org/10.3390/polym14091796
7. Mcsweeney EE, Kropa EL (1955) Plastics. Chem Eng News 33:16–19. https://doi.org/10.1021/cen-v033n001.p016
8. Biscaia HC, Carmo N (2023) Bond assessment between rebars embedded into a parent material using a single-function bond-slip model. Constr Build Mater 397:132396. https://doi.org/10.1016/j.conbuildmat.2023.132396
9. Shan Z, Liang K, Chen L (2023) Bond behavior of helically wound FRP bars with different surface characteristics in fiber-reinforced concrete. J Build Eng 65:105504. https://doi.org/10.1016/j.jobe.2022.105504
10. Aiello MA, Leone M, Pecce M (2007) Bond performances of FRO rebars-reinforced concrete. J Mater Civ Eng 1561:205–213. https://doi.org/10.1061/(ASCE)0899-1561(2007)19
11. Baena M, Torres L, Turon A, Barris C (2009) Experimental study of bond behaviour between concrete and FRP bars using a pull-out test. Compos Part B Eng 40:784–797. https://doi.org/10.1016/j.compositesb.2009.07.003
12. Henin E, Tawadrous R, Morcous G (2019) Effect of surface condition on the bond of Basalt fiber-reinforced polymer bars in concrete _ Elsevier Enhanced Reader.pdf. Constr Build Mater 226:449–458
13. Chaallal O, Benmokrane B (1993) Pullout and bond of glass-fibre rods embedded in concrete and cement grout. Mater Struct 26:167–175. https://doi.org/10.1007/BF02472934
14. Nanni A, De Luca A, Jawaheri Zadeh H (2014) Reinforced concrete with FRP bars: mechanics and design. CRC Press
15. American Concrete Institute (2012) ACI 440.3R: guide test methods for fiber-reinforced polymer (FRP) composites for reinforcing or strengthening concrete and Masonry structures pdf download 28
16. European Committee for Standardization (CEN) (2004) Eurocode 2: design of concrete structures – Part 1-1: general rules and rules for buildings. Brussels
17. Rolland A, Argoul P, Benzarti K, Quiertant M, Chataigner S, Khadour A (2020) Analytical and numerical modeling of the bond behavior between FRP reinforcing bars and concrete. Constr Build Mater 231. https://doi.org/10.1016/j.conbuildmat.2019.117160
18. Cosenza E, Manfredi G, Realfonzo R (1997) Behavior and modeling of bond of FRP rebars to concrete. J Compos Constr 1:40–51. https://doi.org/10.1061/(asce)1090-0268(1997)1:2(40)

Chapter 6
Flexural Behavior of Concrete Beams Reinforced with FRP Rebars

Bruno Pedrosa, Cristiane Lopes, Paweł Zielonka, Szymon Duda, Anna Wybraniec, Tomasz Socha, Arkadiusz Denisiewicz, Krzysztof Kula, José A. F. O. Correia, and Grzegorz Lesiuk

Abstract Fiber reinforced polymer (FRP) rebars are a successful solution for reinforcing concrete structures, especially when exposed to aggressive environments. In addition to their corrosion resistance, these rebars have other characteristics that make their application particularly attractive, such as their high strength-to-weight ratio and electromagnetic neutrality. However, for this solution to go from laboratory application to be applied in large scale in the field, careful investigations must be carried out to assess the behavior of the rebars in real concrete structures. This chapter presents some of the results obtained by the authors using hybrid FRP rebars combining glass and carbon fibers. To evaluate the behavior of the beams reinforced with FRP bars, the displacements, deformations, cracking patterns, and failure of all the beams were analyzed considering North American and Canadian standards. The proposed hybrid rebar solution presented good results.

Keywords FRP rebars · Hybrid solutions · Reinforced concrete · Bending performance

---

B. Pedrosa (✉) · J. A. F. O. Correia
CONSTRUCT & INEGI/LAETA, Faculdade de Engenharia, Universidade do Porto, Porto, Portugal
e-mail: bapedrosa@fe.up.pt

C. Lopes
Department of Structural Engineering, Federal University of Minas Gerais, Belo Horizonte, Brazil

P. Zielonka · S. Duda · A. Wybraniec · G. Lesiuk
Faculty of Mechanical Engineering, Wroclaw University of Science and Technology, Wroclaw, Poland

T. Socha · A. Denisiewicz · K. Kula
Institute of Building, University of Zielona Góra, Zielona Góra, Poland

G. Lesiuk et al. (eds.), *Composite Rebars for Reinforced Concrete Structures*, Structural Integrity 34, https://doi.org/10.1007/978-3-031-96292-9_6

6.1 Introduction

Over the past two decades, laboratory tests have demonstrated that FRP bars can be used successfully and practically as internal reinforcement in concrete structures. The role of industry/university cooperative research became a key aspect in transferring the use of internal FRP reinforcement for concrete from the laboratory to the construction field. Due to differences in physical and mechanical behavior of FRP rebars compared to steel, experimental and numerical approaches to analyze the flexural behavior of concrete structures with FRP bars are required. In this sense, this chapter aims to characterize the bending performance of concrete beams with FRP rebars. New types of FRP rebars were developed and used in experiments aiming to deliver a hybrid solution that can provide good mechanical strength allied with reduced costs.

6.2 FRP Rebars

Fiber reinforced polymer (FRP) rebars are a combination of fibers and resin. The fibers are responsible for the mechanical properties of the rebars and the resin for the environmental resistance and transference of stress between fibers. The main types of bars available on the market are glass (GFRP), basalt (BFRP), aramid (AFRP), and carbon (CFRP) and their mechanical properties vary considerably according to the type of fiber used and even for rebars produced with the same type of fiber [1]. The advantages of FRP rebars include corrosion resistance, non-magnetic behavior, lightweight, and high strength [2–4]. In view of the advantages presented by these materials, they appear to be efficient solutions for structures exposed to deicing salt as for example, marine environment [5].

As there is no standardization for the shape of the bars and surface treatments, they vary considerably in the bars available on the market. Bar shapes include round, solid, square, and hollow. Surface treatments include ribbed, sand coated, wrapped, and fabric texture coated as shown in Fig. 6.1.

The most common manufacturing process used to make FRP rebars is a continuous process called pultrusion. The pultrusion process is the most widely adopted solution to produce FRP bars as it combines consistency and economy [6]. Through this process, the fibers are pulled through guides and impregnated with the resin, making it possible to produce bars of different sizes and shapes.

The mechanical behavior of concrete beams reinforced with FRP bars is different from concrete sections reinforced with steel bars. While the stress–strain diagram of steel shows a plateau due to the yielding of the steel, FRP materials show a linear variation of stresses until failure [7]. Due to the lack of ductility of FRP rebars, the failure of beams reinforced with these bars is brittle, which leads many standards to recommend that the sections be over-reinforced so that failure occurs due to concrete

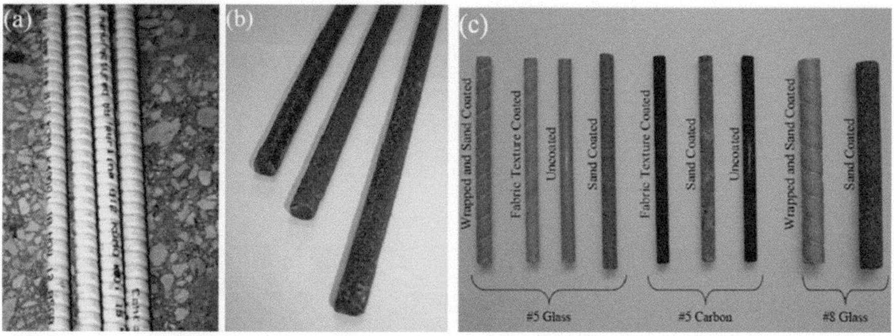

Fig. 6.1 Composite reinforcing rebars: **a** ribbed GFRP rebars, **b** sand coated BFRP rebars and **c** CFRP and GFRP with various surface treatments [1]

crushing and some ductility is added to the beam because of the compressed section of the concrete [8].

Due to the lower modulus of elasticity of most FRP bars when compared to steel as shown in Fig. 6.2, higher deflections and larger crack widths were expected for the beams reinforced with these rebar [9]. Glass fibers are the most commonly used, but due to their low modulus of elasticity they have high deformation values, which limits their application due to the limits of deformability in service. Carbon bars, on the other hand, have a higher modulus of elasticity, but the level of deformation at failure is very small, and they are also more expensive than other types of FRP. Basalt FRP are the most recently developed kind of fibers; it has comparable cost of GFRP and greater chemical stability. Besides that, they are environmentally friendly and have higher strength and modulus of elasticity [10]. Aramid fibers behave in an intermediate range between carbon, glass, and basalt fibers, both in terms of stiffness and strength [11].

The relationship between moment and crack width for beams reinforced with FRP rebars is linear, unlike those reinforced with steel, which show bilinear behavior as shown in Fig. 6.3. As can be seen is the graph above for CFRP rebars (C2-4 to 8) and steel rebars (ST-4 to 6) [7]. It has also been observed [13] that increasing the reinforcement ratio led to a reduction in the width of cracks in beams reinforced with GFRP bars, as well as a significant reduction in crack widths with an increase in the concrete's compressive strength of 52% when the strength was increased from 25 to 45 MPa and 80% when the compressive strength was increased from 25 to 70 MPa.

For FRP-reinforced concrete beams two failure modes can occur, the first for under-reinforced sections, where failure occurs due to rupture of the FRP rebars in tension, and the second for over-reinforced sections where failure occurs due to concrete crushing [14]. Considering the two existing failure modes, failure due to concrete crushing is more desirable because of the increased energy absorption and higher deformability resulting in a more progressive failure mode [6]. The distribution of stress and strain for the two modes of failure, concrete crushing and FRP failure and for the case of balanced section is illustrated in Fig. 6.4.

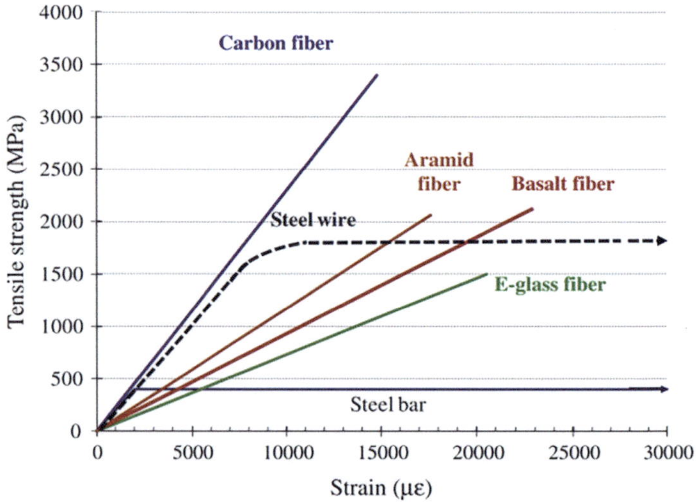

Fig. 6.2 Stress–Strain diagram for FRP and steel bars [12]

Fig. 6.3 Moment crack width relationship [7]

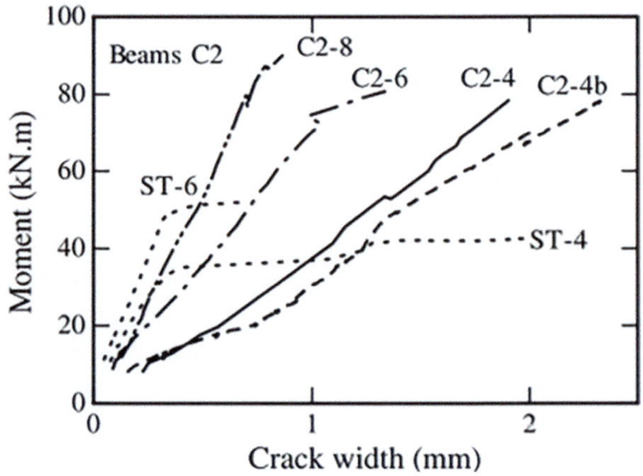

The flexural capacity of concrete structures reinforced with FRP bars is defined considering the principles of internal equilibrium of forces and compatibility of deformations. In the design, the tensile strength of the concrete is disregarded, the perfect bond between the FRP bars and the concrete is considered, and the deformations are proportional to the distance from the neutral axis. The definition of the formulations used in the design depends on the classification of the section as over-reinforced or under-reinforced, this classification being made by comparing the reinforcement ratio with the balanced reinforcement ratio (reinforcement ratio where the two failures would occur simultaneously).

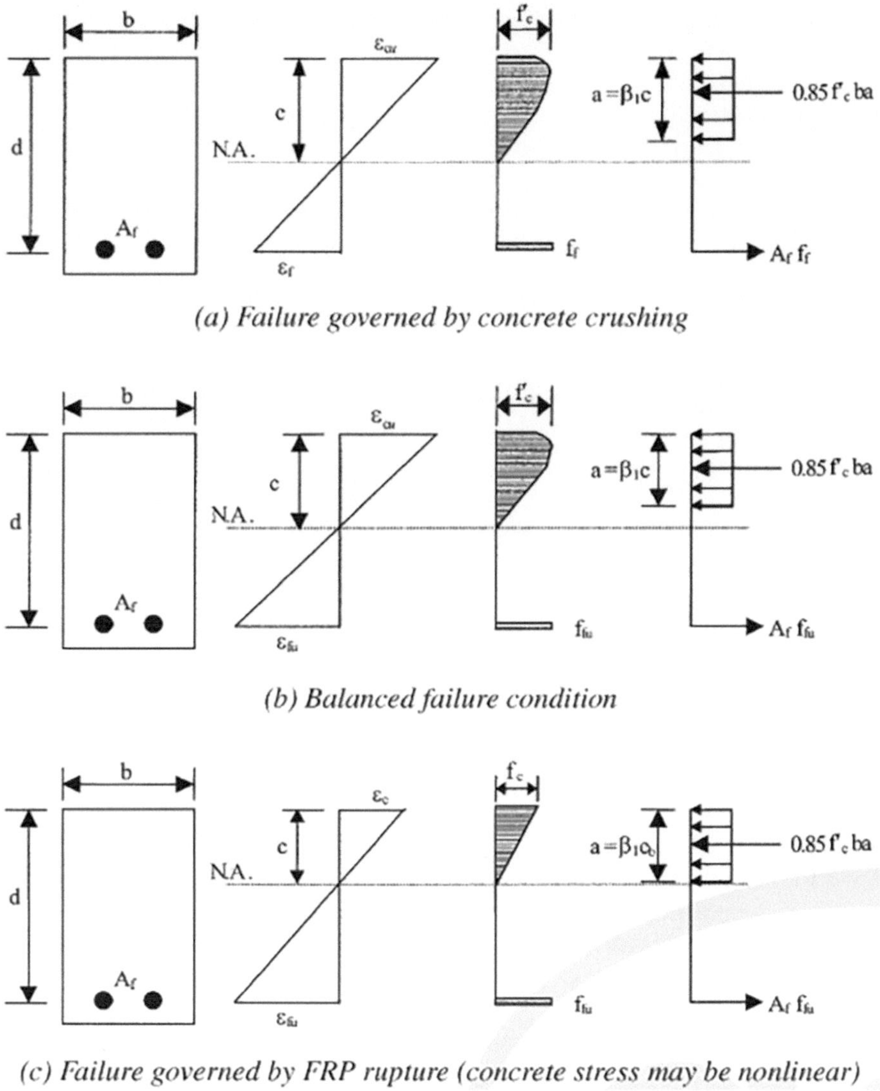

Fig. 6.4 Stress and strain distribution at ultimate conditions [15]

Among the possible solutions to solve the problem of the lack of ductility of FRP bars, in addition to what is suggested by the standard which would be to over-reinforce the section ensuring a certain ductility due to failure by concrete crushing, there are more solutions found in the literature, the combination of steel and FRP bars in the same beam [16, 17], the adoption of hybrid FRP bars [18–21], and using more sensitive inspection methods on the concrete so that the failure can be detected before it happens [22]. The combination of steel bars and FRP as beam reinforcement occurs in such a way that the FRP bars are closer to the outside of the beam, as they are more resistant to corrosion. The steel bars, on the other hand, are inserted further inside the beams, becoming more protected against corrosion, due to the greater coverage, and contributing to the beams' ductility and serviceability.

In the research project "OPTiBREaKuP", the solution tested to overcome the problem of the lack of ductility of FRP bars was the production of hybrid bars. The process of producing this type of FRP bar is described in the literature as hybridization. Through this process a pseudo ductility is acquired by the bar by using different types of fibers.

6.3 Experimental Investigation

The goal of the experimental campaign was to evaluate the effectiveness of developed hybrid FRP rebars as tensile reinforcement of concrete beams under bending. Full-size concrete beams were tested using three different types of reinforcement: commercially available GFRP rebars and two types of developed hybrid rebars. The results presented in this chapter are part of a larger research project conducted with the purpose of developing and testing new hybrid FRP rebars for concrete structures [23, 24].

6.3.1 Materials

The developed new hybrid rebars were produced using the pultrusion process. It allows for fast and low-cost production, and it is a widely used technique for FRP profiles. It comprises pulling fibers through a system of tooling devices, passing through a resin impregnations bath and ending in the curing process that transforms the fibers and resin into a single material [25]. It allows for continuous lengths and constant cross-section [26].

Rebars for concrete structures require a rough surface to produce high levels of adhesion between materials. This rough surface is usually obtained through subsequent surface treatment, as the pultrusion process results in smooth bars. In this experimental campaign, developed hybrid rebars were manufactured with a modified pultrusion process that incorporate the mechanisms to get a surface of rebar that do not require a second production stage to achieve the necessary bonding strength to concrete.

The first rebar to be manufactured was designated as Hybrid55 relating to its fiber content of 55%. A combination of glass fibers (95%) and carbon fibers (5%) was used. This rebar is shown in Fig. 6.5a. The modified pultrusion process used for this rebar is described in Chapter II as the pultrusion process with braiding module (rotating disks) and heat zone. With the addition of these disks, it is possible to produce FRP rebars with a configuration very close to that of the helically wrapped

Fig. 6.5 New hybrid rebars: **a** Hybrid55; **b** Hybrid67

rebars found in literature. However, it should be noted that this production method generated a greater number of voids and the occurrence of packing of the fibers.

The second rebar to be manufactured was designated as Hybrid67 relating to its fiber content of 67%. This bar has the same percentage of glass and carbon fibers as Hybrid 55. Hybrid67 rebar is shown in Fig. 6.5b. The modified pultrusion process used to manufacture Hybrid67 rebars was developed to avoid the problems presented by the previous method. A heated die was added before the rotating disks to consolidate the axial fibers and reduce the content of resin in the rebars. In this kind of pultrusion process, the fibers are pulled into the heated die where the resin is still in the liquid form, then the fibers involved in the resin are pulled out and another bundle of fibers is wrapped around the rebars using the rotating disks, creating surface irregularities that also resemble the helically wrapped surface. Final curing is done in the heated zone, and finally they are pulled and cut to the desired size. This pultrusion process is also described in Chapter II.

The impact of manufacture technique on the microstructure of each rebar type was investigated by means of microscope observations—see Fig. 6.6. The application of heated die before the rotating disks used in the manufacturing process of Hybrid67 rebars had a significant impact on the consolidation of fibers and removing air from the resin leading to a significant reduction of voids in the volume of rebars. Pre-crosslinked matrix with higher viscosity has less tendency to form pores, which have direct impact on the mechanical properties of the rebars.

6.3.2 *Test Setup*

Experimental tests were conducted using a four-point bending setup illustrated in Fig. 6.7. The span was defined as 2300 mm using simply supported conditions. The load was applied in two points whose distance is 500 mm. The cross-section of the beam is defined by 150 × 250 mm—see Fig. 6.8a. This configuration leads to variable bending moments in both sides of the beam and constant value between the load application points (which means no shear load). For reinforced concrete beam specimens using Hybrid67rebars, strain gauges were implemented in two locations at midspan as presented in Fig. 6.7. Tensile strains were measured in Hybrid67 rebars and compression strains were measured at the top surface of the beam as presented in Fig. 6.10.

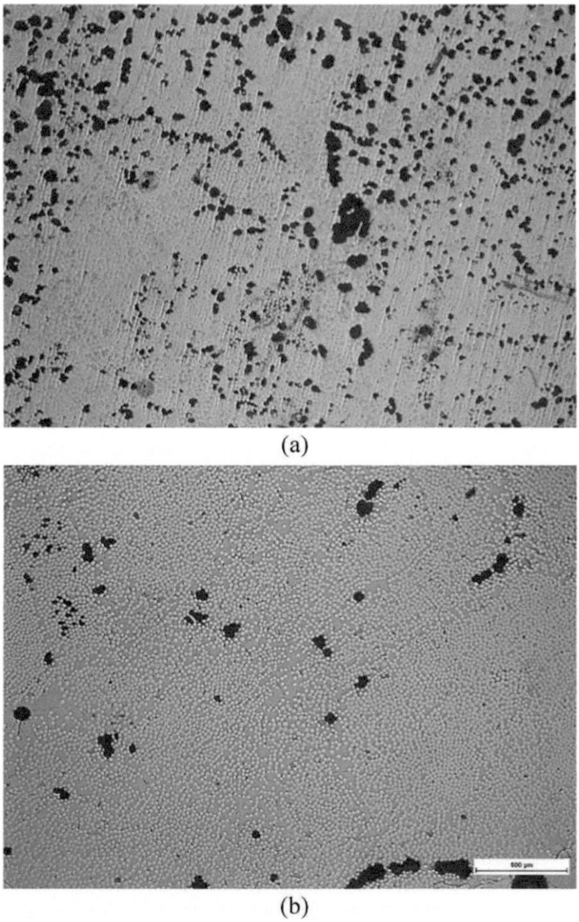

Fig. 6.6 Microscope observations of the composite rebars cross section: **a** Hybrid55 rebars; **b** Hybrid67

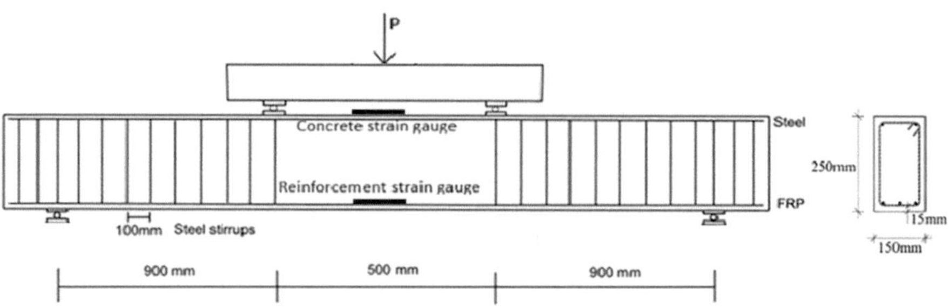

Fig. 6.7 Flexural test setup

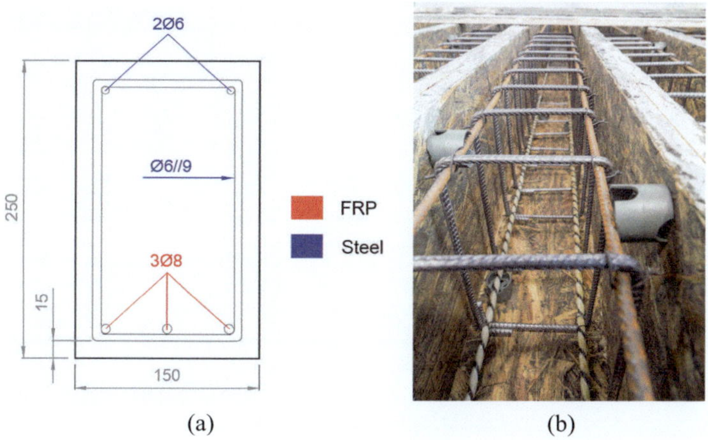

Fig. 6.8 Beam details: **a** schematic representation; **b** image before casting

Figure 6.8b shows the preparation of the rebars before casting. Two steel rebars with diameter of 6 mm were used in the longitudinal direction in the compression zone of the beam and shear reinforcement of the beams was carried out using steel rebars with 6 mm placed with a 9 cm spacing. The shear reinforcement was not applied in the constant moment zone. The FRP rebars were positioned as longitudinal reinforcement in the tension zone of the beam.

There are two potential failure modes for concrete beams reinforced with fiber-reinforced polymer (FRP) rebars under bending loading: either the concrete in the area subjected to compression will rupture, or the rebars in the tensile region of the beam's cross-section will rupture. High-strength concrete was used in the beams (average compressive strength of 67.81 MPa) to avoid failure mode by concrete crushing leading to majority of failure modes by tensile rupture of rebars.

Figure 6.9 presents an FRP-reinforced concrete beam prepared for testing. The load was applied to the beams using an Instron Machine with a load capacity of 500kN as shown in Fig. 6.9a. Deflections were monitored during the test by placing laser sensors in the midspan, as well as LVDTS, which were positioned both in the midspan and at the supports (Fig. 6.9b). The goal was to perform these tests to reach the limit strength capacity of each type of rebar used.

Table 6.1 summarizes the details of the experimental program. The two hybrid rebars developed for this study and commercial green rebars were characterized by their tensile mechanical properties based on experimental tensile tests. Results are presented in Table 6.1.

(a) (b)

Fig. 6.9 FRP-reinforced beam before testing: **a** loading setup; **b** Deflection measurement at midspan

Fig. 6.10 Strain gauges at the upper surface of the beam at mid-span

Table 6.1 Experimental program details

Identification	Number of Sp	Tensile rebars				
		A_s [cm^2]	Type	E_f [GPa]	f_{fu} [MPa]	ε_{fu} [%]
Glass	3	150.8	Commercial	58	1370–1410	2.4
Hybrid55	3	150.8	Developed	32	500	1.6
Hybrid67	3	150.8	Developed	48	1013	2.0

6.4 Analysis of Experimental Results

6.4.1 *Moment—Deflection Performance*

Experimental data obtained from LVDT at mid-span was correlated with the applied load converted in bending moment. Figure 6.11a–c shows the moment-deflection performance of all specimens tested with Glass, Hybrid55 and Hybrid67 rebars. The resultant mean curve for each type of beam is then compared in Fig. 6.11d. Firstly, it can be stated that the level of scatter is not significant for all beam types. The performance is mainly bilinear. In the first part of the test, there is a higher stiffness and no crack is observed. This initial stiffness, $K_{initial}$, was computed for each test and results are presented in Table 6.2. Hybrid 67 presents higher values of initial stiffness when compared to Hybrid55. Even if Glass-rebar beams presented higher values of initial stiffness, the scatter is more relevant than the compared hybrid solutions. The higher level of scatter for Glass-rebar beams is also noticed for the cracking moment, M_{crack}, which marks the transition for the second part of the moment-deflection performance of FRP-reinforced concrete beams. At the beginning of this second stage of the test, all specimens presented several load drops which can be explained by crack opening in concrete. Hybrid55 and Hybrid67-rebar beams presented similar cracking moment—between 5 and 6 kNm. In what regards, the stiffness of the second stage of the test, $K_{pos,crack}$, all beams presented a similar value, around 0.4 kNm. Finally, comparing the obtained values for the maximum bending moment for each beam type, it is observed that Hybrid67-rebar beams shown the highest results in contrast with Glass-rebar beams with a difference of nearly 7 kNm.

Strains at concrete and rebars were monitored during the test for Hybrid67 reinforced beams with strain gauges. Results are presented in Fig. 6.12. Compressive strains were obtained at the upper part of the beam, at midspan, in concrete, while tensile strains were obtained in the tensile rebars at midspan. It is visible that the strain level in compression is much higher when compared to the strain level in tension. In fact, when the cracking moment is achieved, the strain level in rebars increases significantly. The lower level of stiffness of FRP materials induces a significant strain increase in the rebars and consequently in concrete above the plastic line. The strain measurement used in rebars was limited to the capacity of the strain gauge.

6.4.2 *Cracking Analysis*

During the bending test of FRP-reinforcement concrete beams, cracks become visible in concrete surface when M_{crack} is achieved. These cracks emerged from the lower side of the section in the constant flexural moment zone, near to midspan, and propagated vertically (perpendicular to the maximum stress direction induced by pure bending) as shown in Fig. 6.13.

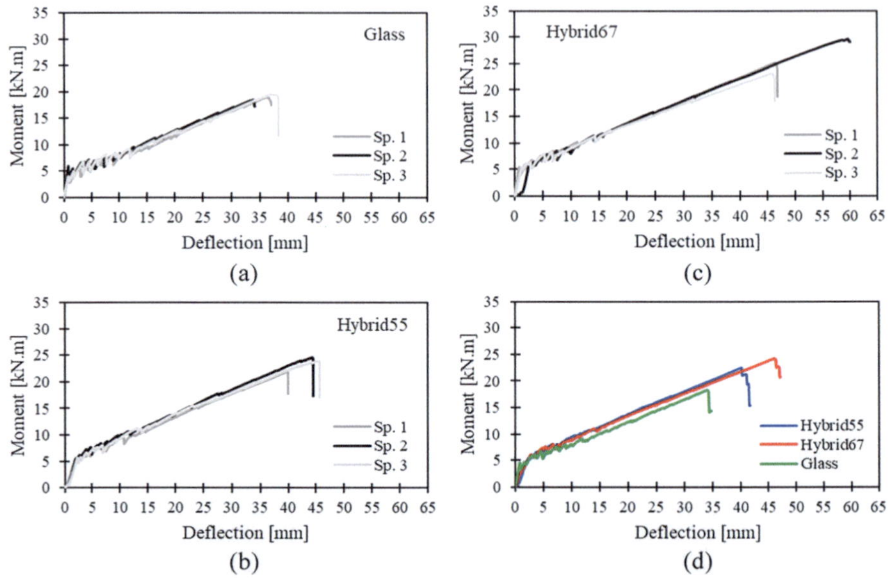

Fig. 6.11 Moment—deflection performance for beams reinforced with fiber reinforced polymer rebars: **a** Glass; **b** Hybrid55; **c** Hybrid67; **d** Comparison of mean curves

Table 6.2 Summary of stiffness and bending moment results from experimental tests

Bart type		$K_{initial}$ [kN m/mm]	M_{crack} [kN m]	$K_{pos,crack}$ [kN m/mm]	M_n [kN m]
Hybrid55	Sp. 1	3.3	4.2	0.44	21.9
	Sp. 2	3.8	5.5	0.45	24.5
	Sp. 3	3.3	5.6	0.43	24.0
	Average	**3.5**	**5.1**	**0.44**	**23.5**
	Std. Dev	0.3	0.8	0.01	1.4
Hybrid67	Sp. 1	6.2	5.3	0.44	25.3
	Sp. 2	4.7	6.0	0.42	29.5
	Sp. 3	4.7	5.4	0.40	23.2
	Average	**5.2**	**5.6**	**0.4**	**26.0**
	Std. Dev	0.8	0.4	0.02	3.2
Glass	Sp. 1	4.7	4.0	0.43	19.1
	Sp. 2	7.7	6.0	0.43	18.5
	Sp. 3	5.2	3.9	0.42	19.6
	Average	**5.9**	**4.6**	**0.4**	**19.1**
	Std. Dev	1.7	1.2	0.01	0.5

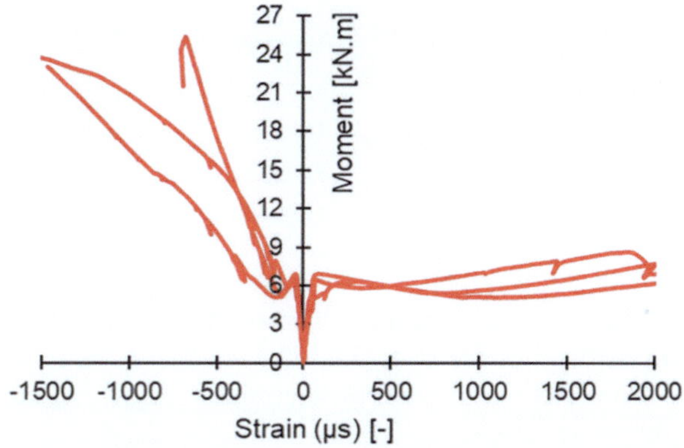

Fig. 6.12 Strain evolution at midspan during bending tests of Hybrid67 reinforced rebars. Compression (negative) strain at top surface of the beam and tensile strains (positive) at FRP reinforcement

Fig. 6.13 Displacement field monitored during bending test using digital image correlation

The performance of each beam until M_{crack} relies on the tensile strength of the concrete which can also be derived from the concrete compressive strength, $f_{c'}$. The value of the cracking moment can be predicted using Eq. 6.1.

$$M_{cr} = \frac{A\sqrt{f_{c'}}I_g}{y_t} \tag{6.1}$$

For constant A, the ACI standard [25] recommends a value of 0.62, while the CSA standard [27] recommends a value of 0.60. The other two parameters of this equation are the gross moment of inertia, $I_g = 1.95\text{E-}4\ \text{m}^4$, and the distance from the centroid axis of gross section, neglecting reinforcement, to tension face, $y_t = 0.125$ m.

Table 6.3 presents the results of the correlation between the experimental and predicted values of the cracking moment. The outcomes of the two criteria were comparable. Predicted values are generally higher than the average cracking moment

Table 6.3 Cracking bending moments: experimental vs predictions

Bar type	$M_{cr,exp}/M_{cr,pre}$	
	ACI	CSA
Hybrid55	0.6	0.7
Hybrid67	0.7	0.7
Glass	0.6	0.6

of FRP beams obtained experimentally, around 30%, due to the difference between the stiffness of the materials. The results showed that it is necessary to correct the code predictions in order to determine the cracking moment for concrete beams with composite rebars.

6.4.3 Ultimate Capacity and Failure Mode Prediction

The uniaxial relationship between the stress and strain of FRP rebar and concrete must be considered in order to design concrete structures reinforced with FRP rebars similarly to steel reinforced concrete structures. The design of concrete structures reinforced with steel rebars is typically calculated considering that the section's failure is controlled by the steel rebars' tensile strength. They present ductile behavior and failure is progressive. Because of their elastic behavior until failure, FRP rebars do not exhibit ductility like steel rebars do. For this reason, a different approach is preferred when designing concrete sections reinforced with FRP rebars. The bending capacity of concrete sections reinforced with FRP rebars depends on whether they are tension-controlled or compression-controlled. The FRP reinforcement ratio must be compared to the balanced reinforcement ratio in order to determine which failure will happen first. Equation (6.2) defines the FRP reinforcement ratio, where the area of concrete is $A_c = 375$ cm^2 and the area of FRP reinforcement is A_f.

$$\rho_f = \frac{A_f}{A_c} \tag{6.2}$$

The balanced FRP reinforcement ratio is computed with Eq. (6.3), where α_1 and β_1 are obtained with Eqs. (6.4) and (6.5) or with Eqs. (6.6) and (6.7) when ACI [25] and CSA [27] standards are used, respectively. f_{fu} is the design tensile strength of FRP defined as the guaranteed tensile strength, f_{fu}^*. E_f is the design of guaranteed modulus of elasticity of FRP. The value for the concrete maximum compressive strain, ε_{cu}, is recommended as 0.3% in ACI and 0.35% in CSA.

$$\rho_{fb} = \alpha_1 \beta_1 \frac{f_c{\prime}}{f_{fu}} \frac{E_f \varepsilon_{cu}}{E_f \varepsilon_{cu} + f_{fu}} \tag{6.3}$$

$$\alpha_1 = 0.85 \tag{6.4}$$

$$\beta_1 = 0.85 - 0.05\frac{f_{c'} - 27.6}{6.9} \tag{6.5}$$

$$\alpha_1 = 0.85 - 0.0015f_{c'} \geq 0.67 \tag{6.6}$$

$$\beta_1 = 0.97 - 0.0025f_{c'} \geq 0.67 \tag{6.7}$$

The balanced ratio is the ratio where the rebars' tensile rupture and the concrete's crushing occur simultaneously. If the reinforcement ratio is lower than the balanced ratio ($\rho_f < \rho_{fb}$), the rupture occurs for the rupture of the FRP rebars in tension. When the reinforced ratio is higher than the balanced ratio ($\rho_f > \rho_{fb}$) the failure will occur because of the crushing of concrete in the compressed section of the beam. Table 6.4 presents the reinforcement ratio of each beam type, and the corresponding balanced reinforcement ratio calculated with ACI [25] and CSA [27], showing similar results. Hybrid67-rebar beams present a reinforcement ratio closer to the balanced value due to its higher ultimate tensile strength when compared to Hybrid55. Figure 6.14 illustrates an example of FRP rebar failure obtained in this experimental program.

In case of failure by tensile rupture of FRP rebars the nominal flexural strength is defined as shown in Eq. (6.8) where $d = 22.5$ cm and c_b is obtained with Eq. (6.9).

$$M_n = A_f f_{fu}\left(d - \frac{\beta_1 c_b}{2}\right) \tag{6.8}$$

Table 6.4 Reinforcement ratio and failure mode for each beam type

Bar type	ρ_f [%]	ρ_f / ρ_{fb}		Failure mode	Axial stiffness $E_f A_f$ [MN]
		ACI	CSA		
Hybrid55	0.45	0.4	0.3	FRP failure	4.8
Hybrid67	0.45	1.0	0.9	FRP failure	7.2

Fig. 6.14 Example of hybrid FRP rebar failure after reinforced concrete four point bending test

Table 6.5 Maximum bending moments: experimental vs predictions

Identification	$M_{n,exp}/M_{n,pre}$	
	ACI	CSA
Hybrid55	1.5	1.5
Hybrid67	0.8	0.8

$$c_b = d\left(\frac{\varepsilon_{cu}}{\varepsilon_{cu} + \varepsilon_{fu}}\right) \tag{6.9}$$

The comparison between experimental and predicted values of the maximum moment is presented in Table 6.5. Predictions represent a very low value when compared with experimental results for Hybrid55, even with a smaller difference. On the other hand, predictions for Hybrid67 are generally in accordance with experimental evidence. CSA predictions show higher values when compared to ACI predictions. However, the difference is insignificant.

The experimental results of Hybrid55 and Hybrid67 rebars are compared to the strength predictions obtained in accordance with the ACI [25] standard in Fig. 6.15. The nominal flexural strength (M_n) is the resistance moment before applying the strength reduction factor Ø. As high-strength concrete was used for all sections, the comparison between the reinforcement ratio and the balanced reinforcement ratio ($\rho_f \leq \rho_{fb}$) shows that the expected failure mode is rupture of the FRP rebars. The values determined using the formulations proposed by the standard are generally lower than those obtained experimentally when the strength reduction factor is considered (equal to 0.55 because of the failure of the section by FRP rupture). The results show that the standard can safely predict the flexural strength of beams reinforced with FRP bars.

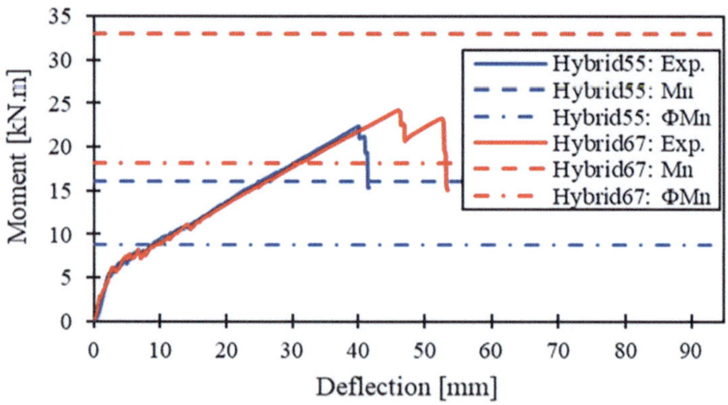

Fig. 6.15 Ultimate capacity predictions based on ACI [25]

6.4.4 Deflection Prediction

The deflection at midspan is defined as shown in Eqs. (6.10) to (6.15) below. It is possible to separate the mid-span deflection of reinforced concrete beams into two phases, an initial phase where the concrete located in the tensile region of the beam does not show cracks and thus the behavior is controlled by the gross moment of inertia (I_g). And a second phase, where the bending moment in the section M_a exceeds the cracking moment M_{cr} and the section begins to show cracks, causing a decrease in the moment of inertia of the section, and this phase is then controlled by this moment of inertia, called the cracked moment of inertia (I_{cr}) [28, 29]. The variation in stiffness along the length of the member is considered with the additional factor γ, as described in ACI and proposed by Bischoff [30, 31]. The deflection prediction was made considering Eq. (6.10) that is derived from linear elastic analysis of simply supported beams subjected to two-point loading.

$$\Delta = \frac{Pa}{48E_cI_e}(3L^2 - 4a^2) \tag{6.10}$$

$$I_e = \frac{I_{cr}}{1 - \gamma\left(\frac{M_{cr}}{M_a}\right)^2\left[1 - \frac{I_{cr}}{I_g}\right]} \leq I_g \tag{6.11}$$

$$k = \sqrt{2\rho_f n_f + \rho_f^2 n_f^2} - \rho_f n_f \tag{6.12}$$

$$n_f = \frac{E_f}{E_c} \tag{6.13}$$

$$I_{cr} = \frac{bd^3k^3}{3} + n_f A_f d^2(1 - k)^2 \tag{6.14}$$

$$\gamma = 1.72 - 0.72\left(\frac{M_{cr}}{M_a}\right) \tag{6.15}$$

The CSA standard [27] disregards the contribution of the cracked section to the stiffness of the beam and determines the deflection of beams through the integration of moment–curvature along the span at any point, as shown in Eqs. (6.20) to (6.22) below.

$$\Delta = \frac{PL^3}{48E_cI_{cr}}\left[3\left(\frac{a}{L}\right) - 4\left(\frac{a}{L}\right)^3 - 8\eta\left(\frac{L_g}{L}\right)^3\right] \tag{6.16}$$

$$\eta = \left(1 - \frac{I_{cr}}{I_g}\right) \tag{6.17}$$

Fig. 6.16 Comparison between experimental and predicted deflections

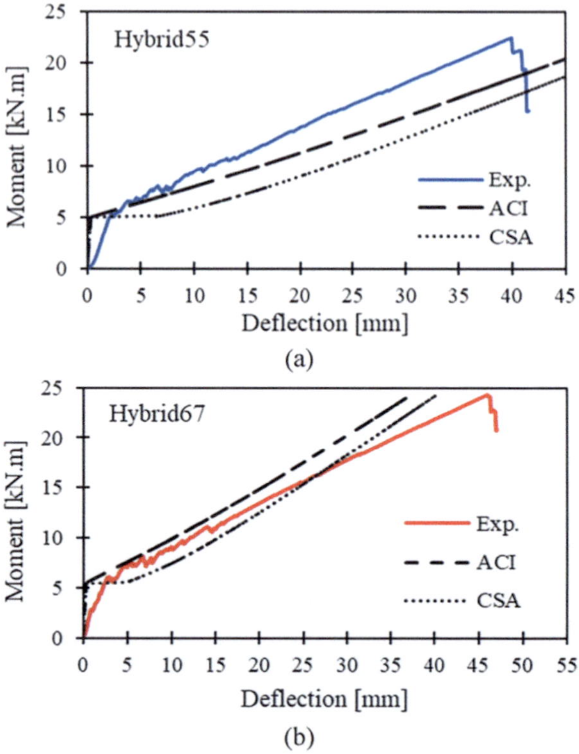

$$L_g = a\frac{M_{cr}}{M_a} \tag{6.18}$$

Figure 6.16 shows a comparison between the experimental deflection results found for Hybrid55 and Hybrid67 and ACI [25] and CSA [27] values. It is consistent that CSA [38] prediction led to higher deflections when compared to ACI [25] predictions. The best correlation between experimental and predicted values is obtained for Hybrid67 beams.

The CSA standard [27] presents more conservative results, overestimating the deflection for all the moment values analyzed.

6.5 Conclusions

FRP reinforced concrete structures have been considered as a valuable solution for many civil engineering applications. However, the mechanical properties of FRP rebars, the adhesion strength between concrete and composite rebars and the manufacturing process of hybrid solutions are still a challenge. This experimental campaign aimed to establish a comparison between developed hybrid FRP rebars with commercially available solutions. For this purpose, full-size concrete beams reinforced with these rebars were built and tested under four-point bending setup. The following conclusions can be addressed:

- Hybrid solutions considering glass and carbon fibers (in a ratio of 95% and 5%, respectively) presented an improved performance when compared to commercially available Glass rebars especially in what concerns the maximum resistant bending moment.
- Strains were measured for Hybrid67 reinforced rebars considering both tension (in rebars) and compression (in concrete). When the cracking moment is achieved, the strain level in rebars increases significantly.
- In terms of crack analysis, both American and Canadian standards presented higher values compared to experimental evidence. Results from tests showed to be 60% of standard values, consistently.
- The two standards also presented very similar values for the maximum bending moment. They proved that the standards can reliably predict the bending resistance by predicting results well below those observed in the beam failure.
- For the deflection predictions, the ACI standard presented results closer to the experimental ones but underestimated the deflections for the sections with the lowest reinforcement ratio. In contrast, the CSA standard presented more conservative results, overestimating the deflection for all the sections analyzed.

Acknowledgements The research was financed by the National Center for Research and Development of Poland (NCBiR) grant number LIDER/40/0219/L-10/18/NCBR/2019.

References

1. Atutis M, Kawashima S (2020) Analysis of flexural concrete beams prestressed with basalt composite bars. Analytical-experimental approach. Compos Struct 243. https://doi.org/10.1016/j.compstruct.2020.112172
2. Wei B, He X, Zhou M, Wang H, He J (2024) Experimental study on flexural behaviors of FRP and steel bars hybrid reinforced concrete beams. Case Stud Constr Mater 20. https://doi.org/10.1016/j.cscm.2023.e02759
3. Timinskas E, Jakštaitė R, Gribniak V, Tamulėnas V, Kaklauskas G (2014) Accuracy analysis of design methods for concrete beams reinforced with fiber reinforced polymer bars/Kompozitais armuotų betoninių elementų projektavimo metodų tikslumo analizė. Eng Struct Technol 5:123–133. https://doi.org/10.3846/2029882x.2013.869415
4. Lau D, Pam HJ (2010) Experimental study of hybrid FRP reinforced concrete beams. Eng Struct 32:3857–3865. https://doi.org/10.1016/j.engstruct.2010.08.028
5. Kara IF, Ashour AF, Köroğlu MA (2015) Flexural behavior of hybrid FRP/steel reinforced concrete beams. Compos Struct 129:111–121. https://doi.org/10.1016/j.compstruct.2015.03.073
6. Bakis1 CE, L. C. Bank FAVLBMAECJFDAMAJJLAMSHRFA and TCTMA (2002) Fiber-reinforced polymer composites for construction—state-of-art review. J Compos Constr 73–87.

7. Kassem C, Farghaly AS, Benmokrane B (2011) Evaluation of flexural behavior and serviceability performance of concrete beams reinforced with FRP bars. J Compos Constr 15:682–695. https://doi.org/10.1061/(asce)cc.1943-5614.0000216
8. Ashour AF (2006) Flexural and shear capacities of concrete beams reinforced with GFRP bars. Constr Build Mater 20:1005–1015. https://doi.org/10.1016/j.conbuildmat.2005.06.023
9. El Refai A, Abed F, Al-Rahmani A (2015) Structural performance and serviceability of concrete beams reinforced with hybrid (GFRP and steel) bars. Constr Build Mater 96:518–529. https://doi.org/10.1016/j.conbuildmat.2015.08.063
10. Wang X, Wu Z (2010) Evaluation of FRP and hybrid FRP cables for super long-span cable-stayed bridges. Compos Struct 92:2582–2590. https://doi.org/10.1016/j.compstruct.2010.01.023
11. Strong ABrent (2008) Fundamentals of composites manufacturing : materials, methods and applications. Society of Manufacturing Engineers.
12. Wang X, Wu Z, Wu G, Zhu H, Zen F (2013) Enhancement of basalt FRP by hybridization for long-span cable-stayed bridge. Compos B Eng 44:184–192. https://doi.org/10.1016/j.compositesb.2012.06.001
13. Adam MA, Said M, Mahmoud AA, Shanour AS (2015) Analytical and experimental flexural behavior of concrete beams reinforced with glass fiber reinforced polymers bars. Constr Build Mater 84:354–366. https://doi.org/10.1016/j.conbuildmat.2015.03.057
14. Akiel MS, El-Maaddawy T, El Refai A (2018) Serviceability and moment redistribution of continuous concrete members reinforced with hybrid steel-BFRP bars. Constr Build Mater 175:672–681. https://doi.org/10.1016/j.conbuildmat.2018.04.202
15. ACI Committee 440.1R-15 (2015) ACI440.1R-15 guide for the design and construction of structural concrete reinforced with firber-reinforced polymer (FRP) Bars, vol 22
16. Zhou B, Wu R, Lu S, Yin S (2021) A general numerical model for predicting the flexural behavior of hybrid FRP-steel reinforced concrete beams. Eng Struct 239. https://doi.org/10.1016/j.engstruct.2021.112293
17. Bencardino F, Condello A, Ombres L (2016) Numerical and analytical modeling of concrete beams with steel, FRP and hybrid FRP-steel reinforcements. Compos Struct 140:53–65. https://doi.org/10.1016/j.compstruct.2015.12.045
18. Saikia B, Thomas J, Ramaswamy A, Nanjunda Rao KS (2005) Performance of hybrid rebars as longitudinal reinforcement in normal strength concrete. Mater Struct 38:857–864. https://doi.org/10.1617/14229
19. Behnam B, Eamon C (2014) Analysis of alternative ductile fiber-reinforced polymer reinforcing bar concepts. J Compos Mater 48:723–733. https://doi.org/10.1177/0021998313477170
20. El-Sayed TA, Erfan AM, Abdelnaby RM, Soliman MK (2022) Flexural behavior of HSC beams reinforced by hybrid GFRP bars with steel wires. Case Stud Constr Mater 16. https://doi.org/10.1016/j.cscm.2022.e01054
21. Ali NM, Wang X, Wu Z (2014) Integrated performance of FRP tendons with fiber hybridization. J Compos Constr 18. https://doi.org/10.1061/(asce)cc.1943-5614.0000427
22. Bakis CE, Nanni A, Terosky JA, Koehler SW (nd) Self-monitoring, pseudo-ductile, hybrid FRP reinforcement rods for concrete applications
23. Lopes C, Lesiuk G, Duda S, Smolnicki M, Pedrosa B, Błażejewski W et al (2024) Computational approach to the bond behaviour between FRP rebars and concrete. In: MS195A - bridging disciplines and scales for sustainable cement and concrete I
24. Lopes C, Carvalho H, Pedrosa B, Duda S, Zielonka P, Smolnicki M et al (2023) Bond behavior between helically wrapped FRP rebars and concrete. In: Numerical modeling of concrete structures
25. American Concrete Institute (2015) ACI 440.1R-15: guide for the design and construction of structural concrete reinforced with fiber-reinforced polymer (FRP) bars
26. Gudonis E, Timinskas E, Gribniak V, Kaklauskas G, Arnautov AK, Tamulėnas V (2014) FRP reinforcement for concrete structures: state-of-the-art review of application and design. Eng Struct Technol 5:147–158. https://doi.org/10.3846/2029882x.2014.889274

27. Canadian Standard Association (2009) CAN/CSA - S806-02: design and construction of building components with fibre-reinforced polymers
28. Yost JR, David W, Dinehart SPG (nd) Effective moment of inertia for glass fiber-reinforced polymer-reinforced concrete beams. ACI Struct J 100. https://doi.org/10.14359/12839
29. Mousavi SR, Esfahani MR (2012) Effective moment of inertia prediction of FRP-reinforced concrete beams based on experimental results. J Compos Constr 16:490–498. https://doi.org/10.1061/(ASCE)CC.1943-5614.0000284
30. Bischoff PH (2005) Reevaluation of deflection prediction for concrete beams reinforced with steel and fiber reinforced polymer bars. J Struct Eng 131:752–767. https://doi.org/10.1061/(ASCE)0733-9445(2005)131:5(752)
31. Bischoff PH, Gross SP (2011) Design approach for calculating deflection of FRP-reinforced concrete. J Compos Constr 15:490–499. https://doi.org/10.1061/(ASCE)CC.1943-5614.0000195

Chapter 7
Dynamic and Cyclic Performance of the FRP Materials

Kayode James Olaleye and Krzysztof Jamroziak

Abstract Large mass or velocity collisions can result in substantial structural damage or collapse even though they are uncommon. They can be brought on, among other things, by wind-generated missiles, car accidents, or weaponry. The con-sequent dynamic inertial loads and the impact of the high strain rates on the material characteristics make it difficult to analyze and design for these effects. Therefore, structural engineering places a high value on any advancements made in impact resistance or dynamic analysis and design processes. For many years, experts have researched how plain and reinforced concrete reacts to impacts. This chapter elaborates dynamic and cyclic performance of FRP material.

Keywords Impact loading · Dynamic behavior

7.1 Introduction

The construction industry's shift to sustainable development depends on the environmentally friendly nature of the building materials used [1–4]. The use of various materials and techniques to increase concrete strength and demonstrate the benefits to the economy and environment has been the subject of numerous review studies. Although historically the aerospace and military industries have been the main users of FRP composite materials, civil engineers are becoming more aware of their benefits [5–8]. Over the past three decades, these materials have become more popular in civil engineering applications due to their superior mechanical properties when compared to typical building materials. FRP materials are corrosion resistant and can be utilized to build new composite members or to reinforce concrete elements that already exist [9–14]. A few advantages of FRP members include faster construction and less expensive maintenance over the course of the structure's life [12–14]. FRP columns have seen an increase in use over the past ten years in the building industry

K. J. Olaleye (✉) · K. Jamroziak
Faculty of Mechanical Engineering, Wroclaw University of Science and Technology, Wroclaw, Poland
e-mail: Kayode.Olaleye@pwr.edu.pl

G. Lesiuk et al. (eds.), *Composite Rebars for Reinforced Concrete Structures*, Structural Integrity 34, https://doi.org/10.1007/978-3-031-96292-9_7

because of their excellent mechanical characteristics [12, 15, 16]. In addition, various fiber materials are being considered as alternatives to conventional steel in reinforced concrete construction due to their continuing cost drop. FRP profiles are frequently employed as structural supporting elements in beam and column applications. There are three different types of FRP tubes used in column applications: (a) FRP tubes, (b) FRP profiles, and (c) hybrid columns comprised of steel, concrete, and FRP tubes [12, 17–19]. The main goal of FRP columns is to confine pressure in the transverse direction of concrete columns by utilizing the strength of the FRP. The second class of FRP profiles, on the other hand, are designed to be used in the production of lightweight structural column components. Due to its affordable construction method, pultruded FRP profiles, which are comparable to modern steel profiles, have recently grown in popularity. Sandwich panels were tested at component and complete scales under a four-point flexural test by Liang et al. [12, 20, 21], and their mechanical responses were subsequently examined using conventional finite element models. They came to the conclusion that CFRP panels were 15–20% lighter, 10–15% stronger, and 50–100% stiffer than GFRP panels. Concrete columns can be reinforced with a variety of materials, including steel, bamboo, CFRP, GFRP, BFRP, AFRP, and others [12].

In the past, impact behavior of composites was investigated using drop weight tests to investigate deflection, load transfer mechanisms, and energy absorption capacity, among other things. These experiments were only conducted on thin plates since the indenter was unable to puncture the target [22–24]. The penetrating resistance of composite laminates has also been the subject of some studies. Experimental research on the behavior of CFRP laminate under bullet impacts at low and high velocities was conducted by Cantwell and Morton. The main energy absorbing mechanisms were found to be perforation, elastic deformation, delamination, and shear out. By using woven S2-glass and graphite fiber as reinforcement in the hybridized composite, Sevkat examined the projectile residual velocity and stresses in the target [25–27]. It was determined that the ballistic limit for sandwich composite was much higher than for single graphite fiber composite. However, research on energy absorption in composite targets from projectile impacts at high speeds was also published in the literature. Experimental analysis of a CFRP composite plate under a low-energy collision with tiny dirt was performed by Bilingardi and Vadori [26]. After studying the relationship between fiber orientation and impactor mass, it was shown that the indentation in the target rises with impactor mass. By employing glass fiber and Kevlar-29 as reinforcement, Muhi et al. investigated the influence of the target's hybridization [28–30].

Ballistic and blast loading have a more brief and dynamic action. A thorough effort is required to protect both civil and military structures by reducing the impact and explosion caused by ballistics through logical material selection and structure design [31–33]. High performing materials like Ultra-High-Performance Concrete (UHPC) are suitable for reducing the impact and blast damage to structures or materials in addition to the enhanced blast resistant design of the structures. These substances should be able to distribute energy by making it more ductile and durable. Bunkers and protective (or sacrificial) walls can be constructed using these materials. Ballistic and blast loading have a more brief and dynamic action [34–36]. A thorough effort is required to protect both civil and military structures by reducing the impact and explosion caused by ballistics through logical material selection and structure design. High performing materials like Ultra-High-Performance Concrete (UHPC) are suitable for reducing the impact and blast damage to structures or materials in addition to the enhanced blast resistant design of the structures. These substances should be able to distribute energy by making it more ductile and durable. Bunkers and protective (or sacrificial) walls can be constructed using these materials [37, 38].

The performance of FRP materials depends on the type of fibers, the matrix material, the manufacturing process, and the environment in which they are used. The key fibers used in FRP include

Glass fibers (GFRP): The most commonly used due to their lower cost and relatively high strength.

Carbon fibers (CFRP): Known for their exceptional stiffness, high tensile strength, and lightweight characteristics.

Aramid fibers (AFRP): Famous for high toughness and impact resistance, used in applications demanding durability.

The polymer matrix, typically a thermoset resin such as epoxy, polyester, or vinyl ester, binds the fibers and transfers loads between them, while protecting them from environmental damage.

7.2 Mechanical Performance

7.2.1 Strength and Stiffness

One of the most significant advantages of FRP materials is their high tensile strength-to-weight ratio. The fibers in FRP composites bear most of the load, and their orientation within the matrix greatly influences the strength and stiffness of the material. FRPs are anisotropic, meaning their properties differ along different directions, which allows engineers to tailor the design based on load requirements.

Tensile Strength: Carbon fiber-reinforced polymers (CFRPs) have an ultimate tensile strength that can exceed 1,500 MPa, much higher than typical structural steels.

Stiffness: The modulus of elasticity for FRPs, particularly CFRP, is significantly higher than that of other materials, providing excellent stiffness, which is crucial in applications where minimal deflection is essential, such as in aerospace components.

7.2.2 *Fatigue and Durability*

FRP composites exhibit superior fatigue resistance compared to traditional materials. For instance, CFRP has a high fatigue life, meaning it can undergo a large number of cyclic loads without failure, making it ideal for aerospace, automotive, and marine applications where cyclic loading is common.

Additionally, FRP's durability is a strong point due to the inert nature of polymer matrices and fibers. GFRP and CFRP are particularly resistant to environmental degradation such as corrosion, making them attractive for infrastructure applications like bridge reinforcement, where exposure to moisture, salt, and chemicals is common.

7.2.3 *Creep and Relaxation*

Creep is the tendency of a material to deform permanently under constant stress over time, while relaxation refers to the reduction of stress under constant strain. FRP composites generally exhibit lower creep compared to conventional materials like steel and aluminum, particularly at room temperature.

For example, CFRP's ability to withstand high stresses without significant deformation over time has made it the material of choice in high-performance applications like aerospace, where long-term dimensional stability is essential.

7.3 Thermal and Environmental Performance

7.3.1 *Thermal Properties*

FRPs exhibit different thermal behaviors depending on the type of fiber and polymer matrix used. In general, carbon fibers have low thermal expansion, which makes CFRP composites dimensionally stable under temperature variations. This is beneficial in precision-critical applications like satellite components or high-performance vehicles.

However, the polymer matrix can be a limiting factor in the thermal performance of FRPs. Most thermoset polymers have relatively low glass transition temperatures (Tg), which means they can soften and degrade at elevated temperatures (typically between 150 and 200 °C). This restricts their use in high-temperature environments unless high-performance matrices like polyimides or phenolic resins are employed.

7.3.2 Environmental Effects

FRPs are highly resistant to corrosion, making them ideal for use in harsh environments, such as marine structures or chemical plants. Unlike metals, which suffer from oxidation and corrosion, FRPs remain stable in saltwater, acidic, or alkaline environments. However, they can suffer degradation when exposed to ultraviolet (UV) radiation over long periods. This can lead to surface discoloration and reduction in mechanical properties, though UV-resistant coatings are often applied to mitigate this issue.

7.4 Performance in Structural Applications

7.4.1 Civil Engineering and Infrastructure

One of the major areas where FRP composites are making significant inroads is in civil engineering. FRPs are being used extensively for retrofitting and strengthening aging infrastructure such as bridges, columns, and beams. GFRP is often chosen for these applications due to its high strength, lightweight nature, and ease of installation compared to traditional materials like steel.

For example, the use of CFRP laminates to strengthen concrete beams has become a popular technique due to their high strength and stiffness. These laminates can be bonded to the surface of concrete structures, significantly increasing the load-carrying capacity without adding substantial weight.

7.4.2 Aerospace and Automotive

The aerospace industry has embraced CFRP materials for decades due to their superior performance in terms of weight savings and high strength. Reducing weight is critical for fuel efficiency and performance, especially in aircraft and spacecraft. CFRP's high stiffness also plays a crucial role in maintaining structural integrity in flight. Notably, advanced aircraft like the Boeing 787 Dreamliner and Airbus A350 XWB use large portions of CFRP in their primary structures, including the wings and fuselage.

In the automotive sector, CFRP is being used in high-performance cars for parts such as body panels, chassis components, and suspension systems. Reducing the vehicle's overall weight leads to improved fuel efficiency and acceleration performance. However, due to its high cost, its use is often limited to luxury and sports cars.

7.4.3 *Marine Industry*

FRP composites, especially GFRP, have been extensively used in the marine industry for the construction of boats, ship hulls, and other marine structures. Their resistance to water absorption and corrosion in seawater makes them an attractive choice over traditional materials like steel and aluminum. Additionally, the lightweight nature of FRP materials improves vessel speed and fuel efficiency.

7.5 Limitations and Challenges

While FRP materials provide numerous advantages, they also come with several challenges like **Costs**: FRP composites, particularly CFRP, are relatively expensive compared to conventional materials. The high cost of carbon fibers and the manufacturing processes involved (such as autoclaving) can limit their use in large-scale construction projects or mass-market products. **Brittleness**: FRPs, especially CFRP, are known to be brittle and may fail suddenly without significant plastic deformation. This is a concern in situations where impact or high-stress conditions may occur unexpectedly. **Complex Manufacturing and Repair**: The manufacturing and repair of fiber-reinforced polymer (FRP) composites in civil engineering are inherently complex processes due to the material's intricate structure and the need for precision to ensure long-term performance. FRP composites consist of reinforcing fibers, such as glass, carbon, or aramid, embedded in a polymer matrix. This combination provides exceptional strength-to-weight ratios, durability, and resistance to environmental degradation, making them invaluable in applications such as bridges, buildings, and protective structures. However, their fabrication and repair require specialized techniques and expertise, which present unique challenges in civil engineering.

The manufacturing of FRP composites involves processes such as pultrusion, filament winding, hand lay-up, and vacuum-assisted resin transfer molding (VARTM). Each method must be carefully selected based on the desired mechanical properties, shape, and application. For instance, pultrusion is ideal for producing long, continuous profiles such as beams, while VARTM is suited for creating complex shapes with consistent fiber-resin distribution. Achieving uniformity in fiber alignment and proper resin impregnation is critical to ensuring structural integrity and performance under load. Variations in curing conditions, such as temperature and pressure, can lead to defects like voids, delamination, or uneven fiber distribution, which compromise the composite's strength and durability. Therefore, strict quality control measures, including non-destructive testing techniques like ultrasound or infrared thermography, are essential during manufacturing.

Repairing FRP composites is equally intricate due to their anisotropic properties and susceptibility to localized damage, such as cracking, delamination, or fiber breakage. Repair strategies must consider the original manufacturing method, material composition, and the nature of the damage. Techniques such as patching, injection of resins, and fiber replacement require precise application to restore the structural integrity of the composite. The interface between the original and repaired sections is particularly critical, as improper bonding can lead to stress concentrations and further failure. Advanced adhesives and resins are often employed to ensure compatibility and durability of the repair.

Environmental factors also play a significant role in both manufacturing and repair. FRP composites are sensitive to ultraviolet radiation, moisture, and temperature fluctuations, which can degrade the matrix and reduce adhesion between fibers and the polymer. Protective coatings and additives are often incorporated during manufacturing to mitigate these effects, but they must also be considered during repairs to maintain the composite's resistance to environmental conditions.

The complexity of FRP composite manufacturing and repair in civil engineering stems from the material's advanced properties and the precision required in their handling. Advances in automation, such as robotic lay-up systems and 3D printing of composites, are helping to address these challenges by improving consistency and reducing labor-intensive processes. Similarly, ongoing research into self-healing composites and smart materials with embedded sensors aims to simplify maintenance and extend the lifespan of FRP structures. Despite the challenges, the versatility and performance benefits of FRP composites make their complex manufacturing and repair processes a worthwhile investment for modern civil engineering applications.

7.6 Recent Advancements and Future Trends

With ongoing research and development, FRP materials continue to evolve. Some of the advancements and future trends include **Improved Matrix Materials**: Researchers are working on developing polymer matrices with higher thermal stability and better environmental resistance, such as advanced thermoplastics.

Nano-reinforced Composites: The incorporation of nanomaterials, such as carbon nanotubes, into the matrix can enhance the mechanical and electrical properties of FRP composites, paving the way for multifunctional materials.

Sustainability: Efforts are being made to produce more sustainable FRP composites using bio-based resins and recycled fibers, addressing the environmental concerns related to the disposal and recyclability of conventional FRP materials.

7.7 Ballistic Materials for Protection Purposes

High-strength, high-modulus fibers were produced in the late 1960s to usher in a new era of protection by manufacturing body armor systems against various projectiles, including tiny missiles. The United States Army pioneered body armor with the Interceptor, made up of an outer tactical vest (OTV) with a hard ceramic insert to stop high-speed rifles and pistols. Throughout the world, military forces and law enforcement authorities wear various types of body armor with stiff ceramic plates. These two forms of material can be mixed to protect various regions of the human body, contingent upon the threat level. While armor with larger inserts is necessary to guard against armor piercing, it also results in excessive weights in the armor system, limiting a soldier's mobility in the field. Soft body armor vests constructed of high-strength/modulus fiber-based woven fabrics are now used by military troops and law enforcement authorities. The next section will provide a brief overview of some of the existing protection fiber materials used in body armor and other ballistic protection solutions.

7.7.1 Fiber Materials

The ballistic-resistance armor structure, in essence, prevents the projectile from penetrating and absorbs its kinetic energy via various ballistic absorbing mechanisms such as deformation, surface damages, tension in primary yarns, tension in secondary yarns, delamination and matrix cracking, and so on. As a result, the most important factors influencing protective material performance are strength, modulus, and elongation at break, projectile deformability, and the velocity of the transverse shock waves in the fiber. The following sections attempted to demonstrate the various fibers used in the ballistic protection sector.

7.8 Mechanisms Influencing Ballistic Material Impact Resistance

Various ballistic materials, ranging from felt to metal and composite, have been utilized in various applications, including armor systems, for a number of decades. This has been expanded on bioinspired materials and biomimetic conditions. Although steel was the material most often employed on the battlefield, new and innovative materials such as fibers, composites, laminates, and ceramics have been widely employed in recent decades to meet the demands of various modern military and other operational, technology-driven applications [39]. As a result, textiles have been employed as ballistic material since World War II, as flak jackets for airplanes, and are now commonly used by military and police personnel. Various research have demonstrated that the overall efficacy of the ballistic performance of the materials against impact is dependent on a variety of criteria. Therefore, the responses of ballistic materials cannot be assessed just on solitary fiber qualities or other criteria; rather, several elements interact to generate a structural response [40]. Various academics and scientists have examined those effects utilizing various materials and techniques. Material qualities, fabric structure, projectile geometry and velocity, far field boundary conditions, numerous plies, and friction are all investigated factors which affect ballistic impact performance. The fiber type, yarn characteristics, and fabric unit area weight all have a significant impact on ballistic impact performance. Other factors influencing ultimate performance include ballistic material areal density, target dimension, target plies numbers, target ply layering sequence, and textile construction, such as woven/nonwoven and 2D/3D fabrics [41, 42]. Furthermore, friction between projectile-yarn, yarn-yarn and filament-filament, bullet speed and geometry, target frame size and clamping pressure, alongside shooting angle have all played a substantial part in the energy absorbing process during ballistic impact.

7.9 Analysis of Ballistic Performance of FRP Reinforced Concrete

Fiber-reinforced polymer (FRP) reinforced concrete has gained significant attention in the construction and defense sectors due to its high performance under extreme loading conditions, including ballistic impacts. FRP materials, such as glass (GFRP), carbon (CFRP), and aramid (AFRP) fibers embedded in a polymer matrix, are increasingly replacing traditional steel reinforcements in concrete structures. This analysis explores the mechanisms, properties, and methods used to assess the ballistic performance of FRP-reinforced concrete.

Concrete, inherently strong in compression but weak in tension, has traditionally been reinforced with steel to improve tensile strength and ductility. However, steel is prone to corrosion and lacks the advanced properties required for modern protective structures. FRP reinforcements offer a lightweight, corrosion-resistant alternative with exceptional tensile strength and energy dissipation characteristics, making them ideal for structures exposed to high-velocity impacts, such as bunkers, safe rooms, and protective barriers. Evaluating their ballistic performance involves understanding energy absorption, fracture mechanics, and the interaction between FRP and concrete under dynamic loads.

7.10 Characteristics of FRP-Reinforced Concrete Under Ballistic Impact

The ballistic performance of FRP-reinforced concrete depends on the properties of both the concrete matrix and the FRP reinforcements. Key factors influencing performance include the material properties of FRP, concrete properties, and the bond strength between FRP and concrete.

FRP materials typically exhibit tensile strengths far exceeding those of steel, with values ranging from 500 MPa for GFRP to over 3,000 MPa for CFRP. Their high elongation capacity enables significant energy absorption, which delays crack propagation in the concrete matrix. Additionally, their lightweight nature reduces the dead load of structures, enhancing their energy dissipation characteristics. The concrete matrix's compressive strength and fracture toughness also play a crucial role in resisting the initial compressive forces and minimizing fragmentation under impact.

The interfacial bond strength between the FRP and concrete matrix is critical for load transfer and structural integrity. This bond ensures that the reinforcement effectively supports the concrete in resisting tensile and shear forces, particularly during high-velocity impacts.

7.10.1 Mechanisms of Ballistic Impact on FRP-Reinforced Concrete

The behavior of FRP-reinforced concrete under ballistic impact involves complex interactions between the projectile and the material, resulting in several failure mechanisms. The projectile penetration creates a localized failure zone characterized by cracking, spalling, and fragmentation. The FRP reinforcement delays the spread of damage by distributing tensile stresses along the fibers.

Delamination may occur in multi-layered FRP reinforcements due to high shear stresses generated during impact. While delamination absorbs energy, it can reduce the overall stiffness of the composite. Shear failure and debonding at the FRP-concrete interface can also occur, reducing the effectiveness of the reinforcement. Addressing these mechanisms through optimized material design and bonding techniques enhances the ballistic performance of FRP-reinforced concrete.

7.10.2 Experimental Studies on Ballistic Performance

Experimental studies play a vital role in evaluating the ballistic performance of FRP-reinforced concrete. High-velocity impact testing involves firing projectiles at concrete specimens reinforced with different types of FRP. Parameters such as projectile velocity, angle of impact, and specimen thickness are varied to assess performance. Observations focus on penetration depth, spall size, and residual velocity of the projectile.

Failure analysis using high-speed cameras and sensors captures the progression of cracks, spalling, and energy dissipation during impact. Post-impact examination of the FRP reinforcements and matrix reveals the extent of damage and energy absorption. Material testing, including tensile and bond tests on FRP materials, provides critical data on their mechanical properties under dynamic loads. Microstructural analysis using scanning electron microscopy (SEM) identifies failure modes at the fiber-matrix interface, offering insights into improving composite performance.

7.10.3 Numerical Modeling of Ballistic Impact

Numerical modeling of ballistic impact on fiber-reinforced polymer (FRP) reinforced concrete is a crucial approach for understanding the dynamic response of this advanced material under high-velocity impacts. This technique allows researchers and engineers to analyze complex phenomena such as stress distribution, energy absorption, crack propagation, and material failure, all of which occur in fractions of a second during ballistic events. Finite element analysis (FEA) is the primary method employed for such simulations, offering detailed insights into the material's behavior while reducing the need for extensive physical testing.

To accurately model ballistic impacts, the unique properties of FRP-reinforced concrete must be incorporated into the simulation. FRP composites are anisotropic materials, meaning their mechanical properties vary based on the direction of the applied load. Numerical models must capture the orthotropic behavior of FRP reinforcements, including their high tensile strength and stiffness along the fiber direction and relatively weaker properties in transverse and shear directions. Similarly, the concrete matrix must be modeled to reflect its brittle nature under tension and its ability to undergo crushing under compressive loads. Constitutive material models, such as damage plasticity models for concrete and strain-rate-dependent models for FRP, are commonly used to represent these behaviors.

The interaction between the projectile and the FRP-reinforced concrete is a critical aspect of numerical modeling. Contact algorithms are used to simulate the high-velocity impact, accounting for frictional forces and material penetration. These interactions are often nonlinear, requiring advanced numerical solvers to handle the complexity of the problem. The projectile's material properties, shape, and velocity are key inputs that significantly influence the simulation results.

Another important factor is the bond between the FRP reinforcement and the concrete matrix. Accurate representation of the interfacial behavior is essential for predicting delamination, debonding, and energy transfer during impact. Cohesive zone models (CZMs) are frequently used to simulate the interface, capturing the progressive separation between the FRP and concrete as stresses increase. This modeling approach provides valuable insights into the role of the bond strength in determining the overall ballistic resistance of the material.

Boundary conditions in the numerical model are also critical to ensuring realistic results. The edges of the FRP-reinforced concrete specimen are often fixed or constrained to replicate the conditions of a structural component in the field. The size of the computational domain must be sufficient to avoid edge effects and ensure that the stress waves generated by the impact are accurately captured.

Validation of numerical models is a vital step in the modeling process. Experimental data from ballistics tests are used to calibrate the model and verify its accuracy. Parameters such as penetration depth, crack patterns, backface deformation, and projectile residual velocity are compared between the simulation and experimental results. A well-validated model can then be used to perform parametric studies, exploring the effects of variables such as FRP type, reinforcement configuration, and concrete grade on ballistic performance.

Advances in numerical modeling techniques, including high-performance computing and machine learning, are further enhancing the accuracy and efficiency of simulations. Adaptive meshing techniques allow for fine resolution around the impact zone while maintaining computational efficiency in less critical regions. Machine learning algorithms are being used to optimize model parameters and predict outcomes based on large datasets, reducing the time required for iterative simulations.

Numerical modeling of ballistic impact on FRP-reinforced concrete is a powerful tool for optimizing material design and performance. By providing detailed insights into the dynamic response of these materials, it enables engineers to develop safer and more efficient protective structures. As computational capabilities continue to improve, numerical modeling will play an increasingly integral role in advancing the understanding and application of FRP-reinforced concrete in high-impact scenarios.

7.10.4 Applications of FRP-Reinforced Concrete in Protective Structures

Fiber-reinforced polymer (FRP)-reinforced concrete has become a cornerstone in the design of protective structures due to its superior mechanical properties, durability, and adaptability to various high-stress environments. The unique characteristics of FRP materials, such as their high tensile strength, corrosion resistance, and lightweight nature, make them an ideal choice for applications where traditional materials like steel fail to meet performance or longevity demands. Protective structures, such as blast-resistant walls, impact barriers, and fortified buildings, benefit significantly from the integration of FRP reinforcements, offering enhanced resilience and safety in extreme conditions.

In military and defense applications, FRP-reinforced concrete is extensively used in the construction of bunkers, blast walls, and protective shelters. These structures are often exposed to high-velocity impacts, explosions, and dynamic loading conditions. The ability of FRP materials to dissipate energy effectively and resist spalling or cracking under such forces ensures the integrity of these critical installations. The non-magnetic and electrically non-conductive nature of FRP also makes it a suitable choice for structures requiring electromagnetic compatibility, such as radar installations or ammunition storage facilities.

In civil infrastructure, FRP-reinforced concrete is applied in protective facades for government buildings, embassies, and high-security facilities. These structures often face the dual challenge of resisting both ballistic threats and environmental degradation. FRP reinforcements, with their exceptional corrosion resistance, contribute to the long-term durability of these facades, even in harsh climates or saline environments. Additionally, FRP-reinforced barriers along highways and transportation hubs provide essential protection against vehicle collisions and accidental impacts, ensuring public safety and reducing maintenance costs associated with conventional materials.

The versatility of FRP-reinforced concrete extends to its use in disaster-prone regions, where structures must withstand seismic forces, hurricanes, or floods. The lightweight nature of FRP materials reduces the overall mass of the reinforced concrete, decreasing inertia during seismic events and minimizing structural damage. In flood-prone areas, the non-corrosive properties of FRP ensure that the reinforcement maintains its strength and functionality despite prolonged exposure to moisture. These attributes make FRP-reinforced concrete a preferred material in the design of resilient infrastructure capable of withstanding natural and human-made threats.

Emerging technologies further enhance the applicability of FRP-reinforced concrete in protective structures. Innovations such as hybrid FRP composites, which combine the benefits of multiple fiber types, and smart materials with embedded sensors for real-time monitoring are revolutionizing the field. These advancements allow for the construction of structures that not only resist damage but also provide critical data on their condition, enabling timely maintenance and extending their service life. As the demand for high-performance, low-maintenance protective structures grow, FRP-reinforced concrete continues to play a pivotal role in ensuring safety and resilience across a wide range of applications.

7.10.5 Advancements in FRP-Reinforced Concrete Technology

Advancements in FRP-reinforced concrete technology have significantly transformed its applications and performance in modern civil engineering, addressing challenges associated with traditional materials and expanding its utility in complex structural environments. Fiber-reinforced polymer (FRP) composites, known for their high strength, lightweight properties, and exceptional corrosion resistance, have seen continuous innovation in material science, manufacturing processes, and integration techniques. These advancements have improved not only the mechanical and environmental performance of FRP-reinforced concrete but also its cost-effectiveness and sustainability.

One of the most notable advancements is the development of hybrid FRP composites, which combine different fiber types, such as glass (GFRP), carbon (CFRP), and basalt (BFRP), to optimize performance characteristics. Hybrid FRPs leverage the tensile strength of carbon fibers, the cost-efficiency of glass fibers, and the thermal stability of basalt fibers, creating reinforcements tailored for specific structural requirements. These combinations enhance the versatility of FRP-reinforced concrete, enabling it to perform efficiently in applications requiring both high durability and affordability.

The integration of nanotechnology has further revolutionized FRP-reinforced concrete technology. By incorporating nanoparticles such as graphene, carbon nanotubes, or nanosilica into the polymer matrix, researchers have achieved substantial improvements in the interfacial bond strength between fibers and the matrix. This enhancement leads to greater load transfer efficiency, improved crack resistance, and increased fatigue life of FRP-reinforced concrete structures. Nanomaterials also impart additional properties, such as self-healing capabilities and enhanced thermal or electrical conductivity, broadening the potential applications of these composites.

Smart FRP materials represent another groundbreaking innovation. These materials integrate embedded sensors that allow real-time monitoring of structural health. By capturing data on stress, strain, and environmental conditions, smart FRP-reinforced concrete systems provide valuable insights into the performance and safety of structures, enabling proactive maintenance and reducing the risk of catastrophic failures. This technology is particularly valuable in critical infrastructure such as bridges, tunnels, and protective structures, where early detection of potential issues is essential.

Advancements in manufacturing processes, such as automated pultrusion, filament winding, and resin transfer molding (RTM), have significantly improved the production efficiency and consistency of FRP composites. These techniques reduce labor costs and material wastage while ensuring uniformity in fiber alignment and resin impregnation, which are critical for achieving optimal mechanical properties. The adoption of 3D printing technologies for FRP-reinforced concrete components has further streamlined production, allowing for the creation of complex geometries and customized designs that were previously unattainable using conventional methods.

In the realm of sustainability, researchers are exploring the use of bio-based resins and recycled fibers to reduce the environmental impact of FRP composites. These eco-friendly alternatives maintain the performance of traditional FRP materials while aligning with global efforts to minimize carbon footprints. Additionally, advancements in recycling technologies now enable the recovery and reuse of fibers and resins from decommissioned FRP components, promoting circular economy practices in the construction industry.

Finally, the development of high-performance coatings and surface treatments for FRP-reinforced concrete has addressed concerns related to environmental degradation. These innovations protect the material from UV radiation, moisture, and chemical exposure, extending the lifespan of FRP-reinforced structures in harsh environments. Combined with design improvements and material innovations, these coatings ensure that FRP-reinforced concrete continues to meet the demanding requirements of modern construction.

In conclusion, advancements in FRP-reinforced concrete technology have expanded its capabilities and applications, making it an indispensable material in contemporary civil engineering. From hybrid and nanotechnology-enhanced composites to smart materials and sustainable manufacturing practices, these innovations continue to push the boundaries of what FRP-reinforced concrete can achieve, paving the way for safer, more durable, and environmentally responsible infrastructure.

7.10.6 Challenges and Future Directions

Despite its numerous advantages, the adoption of FRP-reinforced concrete faces challenges. The high initial cost of FRP materials can limit their use in cost-sensitive projects. Advances in manufacturing processes, such as automated pultrusion, aim to reduce production costs and improve accessibility. The lack of standardized testing protocols and design codes for FRP-reinforced concrete also poses a barrier to widespread adoption. Research into the long-term durability of FRP reinforcements under environmental exposure, such as UV radiation and moisture, is ongoing.

7.11 Conclusions

The analysis of ballistic performance in FRP-reinforced concrete highlights its potential as a revolutionary material for protective and structural applications. Through a combination of experimental and numerical investigations, researchers have demonstrated its superior energy absorption, tensile strength, and resistance to high-velocity impacts. As advancements in material science and manufacturing techniques continue, FRP-reinforced concrete is poised to become a cornerstone of modern protective engineering, addressing the growing demand for resilient and sustainable infrastructure.

Occurrences of dynamic load have an impact on structures. This chapter has examined the promising potential of FRP materials in the field of constructions that resist ballistic impact and modern established techniques that may recreate FRP response under diverse ballistic situations include numerical analysis. However, it is expected that these simulations may pique the curiosity of potential researchers for forecasting the reactions to such dynamic loads when combined with appropriate material models and input parameters.

Acknowledgements The research was financed by the National Center for Research and Development of Poland (NCBiR) grant number LIDER/40/0219/L-10/18/NCBR/2019.

References

1. Lopez-Anido RA, Naik TR (2000) Emerging materials for civil infrastructure: state of the art
2. Ponzo FC, Antonio DC, Nicla L, Nigro D (2021) Experimental estimation of energy dissipated by multistorey post-tensioned timber framed buildings with anti-seismic dissipative devices. Sustain Struct 1
3. O'Shea MD, Bridge RQ (2000) Design of circular thin-walled concrete filled steel tubes. J Struct Eng 126:1295–1303
4. Cheng S, Feng P, Bai Y, Ye LP (2016) Load-strain model for steel-concrete-FRP-concrete columns in axial compression. J Compos Constr 20:04016017
5. Teng JG, Lam L (2004) Behavior and modeling of fiber reinforced polymer-confined concrete. J Struct Eng 130:1713–1723
6. Yu-hao Z, Yu-jie H, Zheng W (2021) Research on dynamic characteristics test of wooden floor structure for gymnasium. China Forest Prod Ind 58
7. Li P, Wu Y-F (2016) Stress–strain behavior of actively and passively confined concrete under cyclic axial load. Compos Struct 149:369–384
8. Xiong M-X, Xiong D-X, Liew JYR (2017) Axial performance of short concrete filled steel tubes with high-and ultra-high-strength materials. Eng Struct 136:494–510
9. Dauletbek A, Li H, Xiong Z, Lorenzo R (2021) A review of mechanical behavior of structural laminated bamboo lumber. Sustain Struct 1
10. Rodsin K (2021) Confinement effects of glass FRP on circular concrete columns made with crushed fired clay bricks as coarse aggregates. Case Stud Construct Mater 15:e00609
11. Cheng X, Wei Y, Nie Y, Wang G, Li G (2021) Compressive behavior of bamboo sheet twining tube-confined concrete columns. Polymers (Basel) 13:4124
12. Draper B, Yee WL, Pedrana A, Kyi KP, Qureshi H, Htay H, et al (2022) Reducing liver disease-related deaths in the Asia-Pacific: the important role of decentralised and non-specialist led hepatitis C treatment for cirrhotic patients. Lancet Regional Health–Western Pacific 20
13. Deskovic N, Meier U, Triantafillou TC (1995) Innovative design of FRP combined with concrete: long-term behavior. J Struct Eng 121:1079–1089
14. Liang R, Stanislawski D, Hota G (2021) Material characterization and structural response under earthquake loads of hakka rammed earth buildings. Sustain Struct 1
15. Corbi O, Baratta A, Corbi I, Tropeano F, Liccardo E (2021) Design issues for smart seismic isolation of structures: past and recent research. Sustain Struct 1:000001
16. Islam MM, Choudhury MSI, Amin A (2016) Dilation effects in FRP-confined square concrete columns using stone, brick, and recycled coarse aggregates. J Compos Constr 20:04015017
17. Dundar C, Erturkmen D, Tokgoz S (2015) Studies on carbon fiber polymer confined slender plain and steel fiber reinforced concrete columns. Eng Struct 102:31–39
18. Jiang T, Wang XM, Zhang WP, Chen GM, Lin ZH (2020) Behavior of FRP-confined recycled brick aggregate concrete under monotonic compression. J Compos Constr 24:04020067
19. Liang R, Hota G (2021) Development and evaluation of load-bearing fiber reinforced polymer composite panel systems with tongue and groove joints. Sustain Struct 1
20. Chen GM, He YH, Jiang T, Lin CJ (2016) Behavior of CFRP-confined recycled aggregate concrete under axial compression. Constr Build Mater 111:85–97
21. Xiao Y (2004) Applications of FRP composites in concrete columns. Adv Struct Eng 7:335–343

22. Cantwell WJ, Morton J (1989) Comparison of the low and high velocity impact response of CFRP. Composites 20:545–551
23. Jordan JB, Naito CJ (2014) An experimental investigation of the effect of nose shape on fragments penetrating GFRP. Int J Impact Eng 63:63–71
24. Sevkat E (2012) Experimental and numerical approaches for estimating ballistic limit velocities of woven composite beams. Int J Impact Eng 45:16–27
25. Cantwell WJ, Morton J (1990) Impact perforation of carbon fibre reinforced plastic. Compos Sci Technol 38:119–141
26. Belingardi G, Vadori R (2003) Influence of the laminate thickness in low velocity impact behavior of composite material plate. Compos Struct 61:27–38
27. Muhi RJ, Najim F, de Moura MFSF (2009) The effect of hybridization on the GFRP behavior under high velocity impact. Compos B Eng 40:798–803
28. Tiberkak R, Bachene M, Rechak S, Necib B (2008) Damage prediction in composite plates subjected to low velocity impact. Compos Struct 83:73–82. https://doi.org/10.1016/j.compstruct.2007.03.007
29. Zhang D, Sun Y, Chen L, Pan N (2013) A comparative study on low-velocity impact response of fabric composite laminates. Mater Des 50:750–756
30. Goda I (2022) Defence technology
31. Perumal R, Nagamani K (2014) Impact characteristics of high-performance steel fiber reinforced concrete under repeated dynamic loading. Int J Civil Eng 12:513–520
32. Tabatabaei ZS, Volz JS, Baird J, Gliha BP, Keener DI (2013) Experimental and numerical analyses of long carbon fiber reinforced concrete panels exposed to blast loading. Int J Impact Eng 57:70–80
33. Máca P, Sovják R (2013) Resistance of ultra high performance fibre reinforced concrete to projectile impact. Struct Under Shock Impact XII 126:261
34. Li QM, Reid SR, Wen HM, Telford AR (2005) Local impact effects of hard missiles on concrete targets. Int J Impact Eng 32:224–284
35. Buttignol TET, Sousa J, Bittencourt TN (2017) Ultra high-performance fiber-reinforced concrete (UHPFRC): a review of material properties and design procedures. Revista IBRACON de Estruturas e Materiais 10:957–971
36. Kennedy RP (1976) A review of procedures for the analysis and design of concrete structures to resist missile impact effects. Nucl Eng Des 37:183–203
37. Drdlová M, Buchar J, Krátký J, Řídký R (2015) Blast resistance characteristics of concrete with different types of fibre reinforcement. Struct Concr 16:508–517
38. Tai YS (2009) Flat ended projectile penetrating ultra-high strength concrete plate target. Theoret Appl Fract Mech 51:117–128
39. Kurzawa A, Pyka D, Jamroziak K, Bajkowski M, Bocian M, Magier M et al (2020) Assessment of the impact resistance of a composite material with EN AW-7075 matrix reinforced with α-Al_2O_3 particles using a 7.62 × 39 mm projectile. Materials 13. https://doi.org/10.3390/ma13030769
40. Almusallam TH, Siddiqui NA, Iqbal RA, Abbas H (2013) Response of hybrid-fiber reinforced concrete slabs to hard projectile impact. Int J Impact Eng 58:17–30. https://doi.org/10.1016/j.ijimpeng.2013.02.005
41. Kennedy RP (1976) A review of procedures for the analysis and design of concrete structures to resist missile impact effects. Nucl Eng Des 37:183–203. https://doi.org/10.1016/0029-5493(76)90015-7
42. Li QM, Reid SR, Wen HM, Telford AR (2005) Local impact effects of hard missiles on concrete targets. Int J Impact Eng 32:224–284. https://doi.org/10.1016/j.ijimpeng.2005.04.005

Chapter 8
Circular Economy Frameworks, Composite Materials and Multi-scale Homogenization Theories: State of the Art

Aditya Vidwans, José A. F. O. Correia, Nicholas Fantuzzi, and Patrizia Trovalusci

Abstract The world has faced numerous environmental degradation and resource depletion difficulties because of conventional economic and developmental paradigms. In this context, the circular economy emerges as a useful framework for transition towards sustainable development. Researchers and institutions have been investigating the circular economy paradigm extensively as a potential way to improve the sustainability of our economic system, increase resource efficiency and support low-carbon development strategies to balance industrial expansion with environmental and human health. Composites have become popular in civil engineering and structures due to their several advantages, such as lightweight, high strength, resistance to crack propagation and chemical reactions, corrosion resistance, high impact strength, sustainability, durability, flexibility and long life. Because of these benefits, usage of composites in engineering has grown which has increased need for a simple, effective, precise and workable technique for analysing them. Two separate techniques that are commonly used in numerical frameworks for the analysis of composite materials are macro-mechanical analysis and micro-mechanical analysis. Each of these approaches has its own limitations and disadvantages. As a result, many multi-scale approaches that consider linked analysis at the macro and micro-scales are used for composite analysis. This study aims to present a concise overview of the state-of-the-art related to the circular economy and composite and recyclable material that support it. Furthermore, it introduces homogenization theories and multi-scale approaches necessary to evaluate composite behaviour.

A. Vidwans (✉) · P. Trovalusci
Department of Structural and Geotechnical Engineering, Sapienza University of Rome, Rome, Italy
e-mail: aditya.vidwans@uniroma1.it

J. A. F. O. Correia
INEGI & CONSTRUCT, Faculty of Engineering, University of Porto, Porto, Portugal

N. Fantuzzi
Department of Civil, Chemical, Environmental, and Materials Engineering, University of Bologna, Bologna, Italy

G. Lesiuk et al. (eds.), *Composite Rebars for Reinforced Concrete Structures*,
Structural Integrity 34, https://doi.org/10.1007/978-3-031-96292-9_8

Keywords Circular economy · Composites materials · Multiscale approaches

8.1 Circular Economy

There are 7.7 billion people on Earth, and during the past few decades, their needs and desires have steadily grown. Thus, to guarantee resource availability, sustainable development is essential. There has been an overrun on the Earth's planetary boundaries since the current system of manufacturing and utilization requires the equivalent of 1.7 Earths to meet demand and assimilate the associated waste produced [1]. It is crucial to use Earth's natural resources as efficiently as possible because they are finite [2]. The origins of the Circular Economy (CE) concept can be traced to the 1970s, with its foundational principle of reducing material consumption in manufacturing processes. However, any resource can be used with this idea [3].

With its roots in sustainability principles, the 'Circular Economy' is an innovative approach for the most effective and sustainable use of resources [4]. Utilization of resources at a rate lot more than the ability of Earth to replenish them in terms of soil, water, air, climate and biodiversity has resulted into permanent damages, which can be prevented with the help of CE [5]. The CE concept suggests altering the linear economy's extraction-production-disposal paradigm, which is presently used on huge scale in many industries, in order to make human activities more robust [6]. The model of circular economy put out by the European Parliament is explained in Fig. 8.1.

Turner and Pearce worked on study called 'Sustainable Economic Development,' which examined the relationship between the environment & the economy and introduced the term 'Circular Economy.' To incorporate the concept of intergenerational utility, the study modified the traditional economic paradigm, which is founded on the utilitarian benefit–cost principle [8]. The circular economy concept suggests a sustainable counterpoint to the conventional take-make-use-dispose linear model by emphasizing make-use-reuse-remake recycle. The CE can be described as a framework built on business models that eliminate the traditional 'end-of-life' stage for products. Instead, it prioritizes recovery, reuse, reduction and recycling of materials across all phases of production, consumption and distribution. This approach aims to achieve sustainable development and will result in better environmental quality, social equity and economic prosperity, benefiting both present and succeeding generations. This circular economy model is applicable at three different scales: macro (region, city, country and beyond), meso (eco-industrial parks) and micro (businesses, products and consumers) [9].

The rate or percentage of four strategies, recycling, remanufacturing/refurbishment, reuse/redistributing and maintaining/extending product lifespan, that are applied across different industries and businesses is depicted in Fig. 8.2.

These strategies, which are illustrated in Fig. 8.2, are briefly introduced as follows:

Recycle. Recycling is the process of gathering, sorting, processing and using waste or discarded resources to create new materials. Recycling is extensively used not only

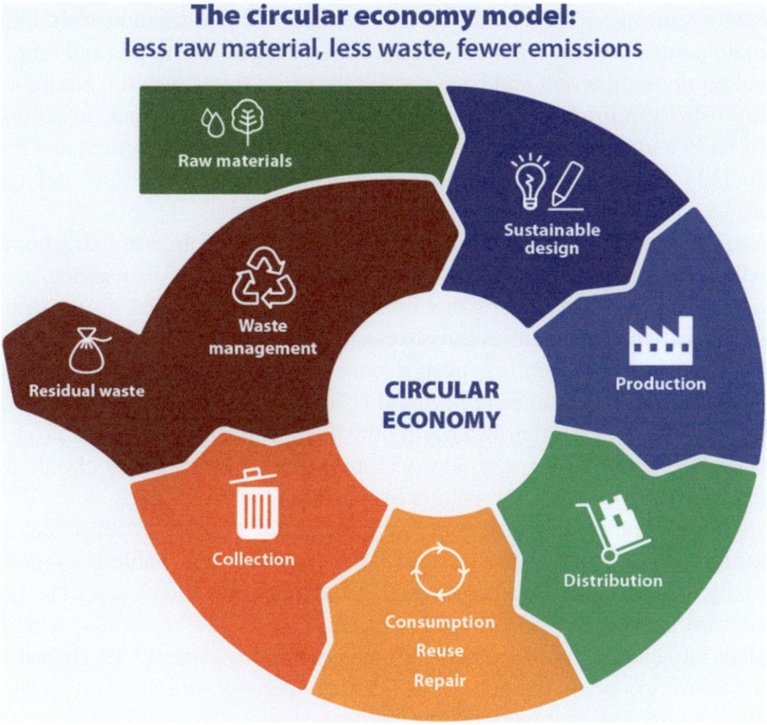

Fig. 8.1 Circular economy model (European Parliament Research Services) [7]

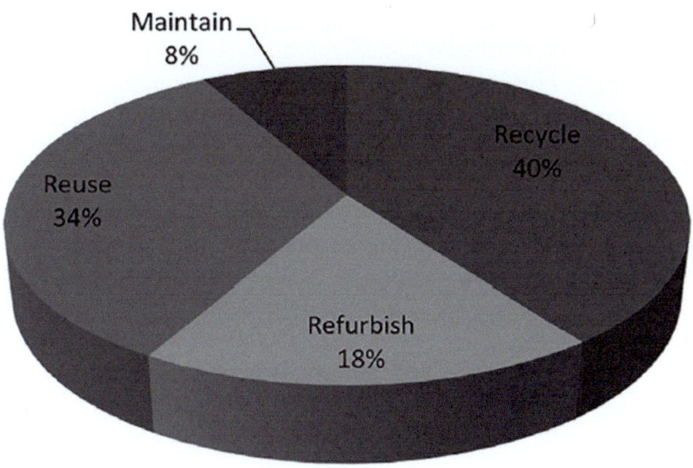

Fig. 8.2 Adoption rate (%) of CE strategies for technical nutrients [10]

in the manufacturing sector overall, but also in a number of other industries, including waste management, textiles, construction, rubber and plastic, metals and fabrication, electrical equipment, wood and paper, machinery and transportation equipment;

Refurbish/Remanufacture. Process of bringing products back to almost new levels of technical performance and quality is known as refurbishment and remanufacturing [11]. Sorting, inspecting, dismantling, cleaning, reprocessing and reassembling are the steps that are involved in this process [12];

Reuse/Redistribute. In simple terms, reusing is the practice of using something repeatedly rather than disposing it away or recycling it [13]. For instance, electrical appliances, food containers, construction formwork, and furnishings are among the many items that can be redistributed or reused; and,

Maintain/Prolong. Creating design concepts that will enhance a product's lifespan and durability, which is necessary for maintaining or prolonging its use. This can be achieved by either modifying the design specifications of a product and researching the potential for upgrades to future models or by using classic designs that guarantee its utilization for as long as feasible [14].

By 2050, population on earth is predicted to reach 9 billion, with severe anthropogenic and environmental repercussions. These repercussions include a reduction in biodiversity, which exacerbates ecological imbalances on a huge scale. The growing demand from millions of people leads to a scarcity of raw materials, which escalates prices and causes volatility in the global economic system [15]. Therefore, this circular economy concept has achieved widespread recognition in many government public policies, including those from the European Union (House of Commons 2014), to address issues of pollution, population expansion, depletion of natural resources and sustainable development [16].

According to the United Nations Environmental Assembly [17], the application of CE can help preserve the worth of resources, products and materials for a considerable duration. The primary objective of a CE is to replicate circularity of nature. Through value-retaining practices like repair, reuse, refurbish, recycle, refuse, rethink and repurpose, the CE converts the existing economic paradigm into a sustainable one. The report by Ellen MacArthur Foundation, published in 2019, suggests that the implementation of CE model in only five essential industries—food, cement, plastics, aluminium and steel—could have a significant environmental impact. This could result in reduction of emissions from goods production by approximately 9.3 billion metric tonnes CO_2 equivalent by 2050 which would account for nearly half of all emissions [18]. This is the same as lowering all transportation-related emissions for 2019 to zero. According to the 2018 IRA Report, 50% of the urban environment needed by 2050 has not yet been constructed [19]. Hence, the environment will be greatly influenced by the use of recycled materials and improved construction methods [20].

The fundamental idea of the circular economy and the several phases that need to be taken into account while putting this idea into practice are illustrated in Fig. 8.3. It can be observed from Fig. 8.3 that the circular economy concept will be implemented throughout the entire process, from design to disposal. A brief synopsis is provided as follows:

Fig. 8.3 Circular economy concept (Source: Marichova, [21]). Licensed under CC BY-NC-ND 4.0.

1. Throughout the design phase, the structure should be intelligently designed to meet all structural criteria with the application of minimum quantity of material. An attempt should be made to design slender structures;
2. An effective manufacturing technique should be used during production. Extra care should be taken during this stage of production and manufacturing to minimize waste generation;
3. Efficient distribution and transportation methods should be searched for;
4. The material must be recycled or reused when the product is no longer considered functional. The reuse of materials in their current configuration should receive more attention because it will minimize the need for modifications, which will save energy; and,
5. Lastly, every feasible recycling technique or idea should be used to reduce waste generation.

8.2 Composites, Polymers and Recyclable Materials

Numerous different industries have expressed interest in composite structures due to their capabilities to provide a superior strength-to-weight ratio to that of non-reinforced materials. Composite materials are made up of two or more components that have been blended in a certain way to keep the constituent materials unique and distinguishable. The primary goal of composites is to mitigate the shortcomings of individual constituent materials while enabling the new material to retain the strengths inherent in its constituent materials. Depending on the composition, processing, shaping and forming methods of the raw materials, composite materials can be precisely customized to the intended use.

The characteristics of its individual components needs to be taken into consideration when developing a composite material for a particular type of application. Strength, toughness and stiffness (modulus of elasticity) are important properties that should be taken into account. Density is quite important in many cases since a weight of a composite may be extremely significant. Furthermore, because composites may be exposed to temperature variations during manufacturing and/or during their in-service life, thermal properties like conductivity and expansion should also be considered. A discrepancy in the thermal expansion coefficients among the constituents may cause internal residual stress to be generated, which could significantly affect the mechanical behaviour of the composite. The properties of several matrix types, reinforcements, certain frequently used industrial materials, and some illustrative composite materials are listed in Table 8.1. According to this table, composites can be used to create advantageous property combinations [22].

A composite is made up of two or more chemically distinct phases that are delineated by a unique interface at the microscopic scale. One component of the composite that is typically incorporated into the matrix and serves to reinforce the composite is called reinforcement. Flake reinforcement, filler reinforcement, particulate reinforcement, long-fibre or continuous-fibre reinforcement, and short-fibre or random-fibre reinforcement are frequently utilized reinforcements. The continuous element, which is typically (though not necessarily) present in larger quantities in composites is called a matrix. In general, addition of another component improves the matrix properties [22]. Composites are often classified into three groups according to the matrix-forming material: Ceramic Matrix Composites (CMC), Metal Matrix Composites (MMC) and Polymer Matrix Composites (PMC). These three composite materials are presented in Fig. 8.4 together with their typical used reinforcement configuration (i.e. fibre reinforcement) and frequently utilized materials [23].

The thermoset or thermoplastic matrix resins, along with fibres that are significantly stiffer and stronger than the matrix, are used in Polymer Matrix Composites (PMC). Compared to unreinforced polymers or traditional metals, polymer matrix composites have the advantages of being stiffer, lighter and stronger. Furthermore, characteristics and shapes of polymer matrix composites can be modified to meet the specifications required for particular applications. High-performance fibre reinforcements, such as carbon fibres, liquid crystalline polymers, organic fibres like aramids and ultrahigh-molecular-weight polyethylene, are particularly interesting for use in aerospace and military applications [22].

Helicopters, aircraft and spacecraft all utilize Metal Matrix Composites (MMC) as structural materials because of their lightweight. Hard reinforcing particles are embedded in a metal matrix phase to create metal matrix composites. Metal alloys with low densities, such aluminium, titanium and magnesium, are frequently utilized as matrix in these composites. It is possible to use nickel super-alloys as a matrix for

Table 8.1 Summary of properties for different material classes [22]

Type of materials	Density (g/cm^3)	Young modulus (GPa)	Tensile strength (Mpa)	Fracture toughness (MPa$\sqrt{m}$)	Thermal conductivity (W/mK)	Thermal expansivity (10^{-6}/K)
Epoxy	1.25	3.5	50	0.5	0.3	60
Nylon	1.1	2.5	80	4	0.2	80
Polyurethane	1.2	0.01	20	0.1	0.2	200
Mild steel	7.8	208	400	140	60	17
Concrete	2.4	40	20	0.2	2	12
Alumina	3.9	380	500	4	25	8
Wood (load// grain)	0..6	16	80	6	0.5	3
(Spruce) (load⊥grain)	0.6	1	2	0.5	0.3	10
General polymer matrix composites (in-plane)	1.8	20	300	40	8	20
Adv. polymer matrix composites (load//fibres)	1.6	200	1500	40	200	0
(APC-2) (load⊥fibres)		3	50	5	40	30
Metal matrix composite (Al-20% SiC_p)	2.8	90	500	15	140	18

Fig. 8.4 Types of composites [23]

high-temperature applications. As continuous fibre reinforcement, boron, carbon and silicon carbide are frequently utilized and scattered throughout the matrix. Alumina, boron carbide and silicon carbide are a few common particle reinforcements. Metal matrix composites have a lower coefficient of thermal expansion, a higher elastic modulus and strength, and greater resistance to creep and rupture [24].

While maintaining additional characteristics including low weight, high stiffness, thermal stability, corrosion resistance and wear resistance, Ceramic Matrix Composites (CMC) are used to improve the toughness and tensile strength of conventional ceramic materials. In ceramic matrix composites, the ceramic matrix phase is reinforced with ceramic whiskers or fibres. The toughness and braking strength of ceramics are significantly increased by whiskers and reinforcing fibres. Strength and toughness can be increased by 100% when ceramics are reinforced with continuous fibres [24]. Table 8.2 provides a general overview of engineering materials, detailing bonds present in the material, their microstructure, advantages and disadvantages.

One of the most popular and widely utilized materials nowadays is synthetic polymers, which are used in manufacturing of high-end technical products like implantable devices as well as for many everyday single-use items like coffee cups and water bottles. As a result, over the past 50 years, there has been a notable growth or surge in both plastic manufacturing and use. Hence, waste generation has become unsustainable and uncontrollable resulting in significant environmental risk [25]. Because of the massive amount of toxic waste produced annually by these non-biodegradable plastics, pollution and land depletion have occurred. These are the

Table 8.2 General characterization of engineering materials [83]

Engineering materials	Bonding	Microstructure	Advantages	Disadvantage
Metals	Metallic	Crystallites	• High strength and stiffness • Good ductility • Conductive	• Fracture • Fatigue
Polymers	Mixed covalent-secondary	Chain molecules	• Cost-effective • Lightweight • Corrosion resistant	• Lower strength • Lower stiffness • Susceptible to creep
Ceramics	Mixed ionic and covalent	Crystal grins, amorphous	• High strength, stiffness and hardness • Temperature resistant • Corrosion resistant	• Prone to brittle fracture
Composites	Multiple types	Matrix and fibre, etc.	• Stiff, strong • Lightweight	• Higher cost • Susceptibility to delamination

greatest dangers facing both the contemporary world and succeeding generations [26, 27].

Presently, very little amount of plastic is recycled, the majority of garbage is either incinerated or disposed of in landfills. Merely 3.14 million metric tonnes, or 9%, of the 34.5 million metric tonnes of discarded plastics/plastic debris produced in the United States in year 2015 were recycled, with the other 91% of the debris being disposed of in landfills [28]. Because of its low density, plastic takes up a lot more space than metals, and it can remain in landfills for years or even decades because it is non-biodegradable, rendering such landfills unsuitable for other purposes. Contrary to this, biodegradable materials can be composted to produce beneficial organic soil in less than a year [29].

The primary advantage of the composites is their superior engineering qualities, which include high stiffness, high tensile strength, weather resistance and lightweight. Many sectors are employing these composites instead of more traditional materials like metals, wood, etc. for precisely these reasons [26]. For instance, polymers like polypropylene and its composites exhibit a number of intriguing mechanical characteristics, including high stiffness, significant elastic limit, lightweight and excellent ageing resistance [30]. The automotive industry makes extensive use of these composites to reduce weight and save money [31]. The raw materials used to manufacture conventional polymer composites are petrochemicals, which are not sustainable. Thus presently, there is a growing need to create sustainable materials that are easy to reuse or dispose of without causing environmental harm. Industrial material producers are now considering health concerns like neurotoxicity linked to conventional materials (non-sustainable), alongside ecological challenges such as the greenhouse effect, rising CO_2 emissions, global climate change and resource depletion [26].

Hence, ideas like material recycling and green composite are becoming increasingly significant. Green composites are composites derived from sustainable and biodegradable polymers that have structural characteristics comparable to those of conventional polymers [32]. The main difference between synthetic and biopolymers is how they behave after the end of life. Because biopolymers are biodegradable, they can break down naturally once their functional life cycle is complete, which contributes to environmental sustainability. Synthetic polymers, on the other hand, usually remain in the environment and either end up in landfills or are burned, which can cause serious environmental pollution. Figure 8.5 provides an illustration of this basic discrepancy.

The examples of synthetic and natural fibres are depicted in Fig. 8.6.

Green composites, or composites made from plant-based ‘green’ polymers and resins, are thus becoming the primary focus of research and will only increase in the future. Owing to considerable challenges like elevated manufacturing costs, difficulties with waste disposal, insufficient recycling methods, high carbon footprint and finite petroleum reserves, researchers are concentrating on application of green composites instead of traditional petroleum-based composites [35, 36].

In order to replace synthetic fibres in real-life applications, an extensive amount of research has been done on resins, such as epoxidized natural rubber [37], linseed

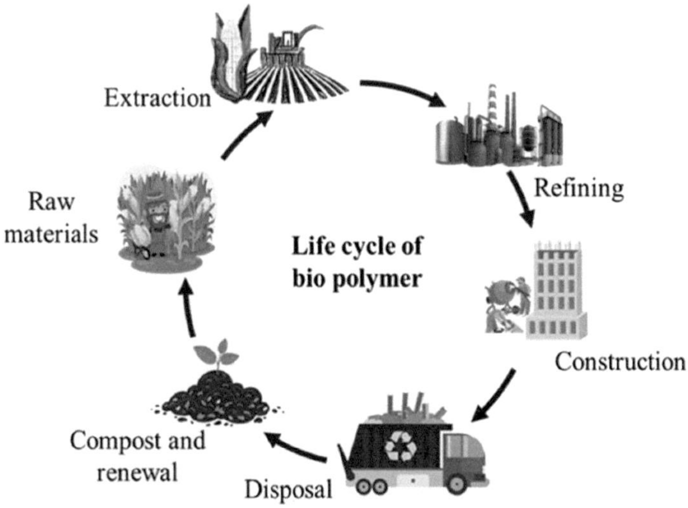

(a) Biopolymer

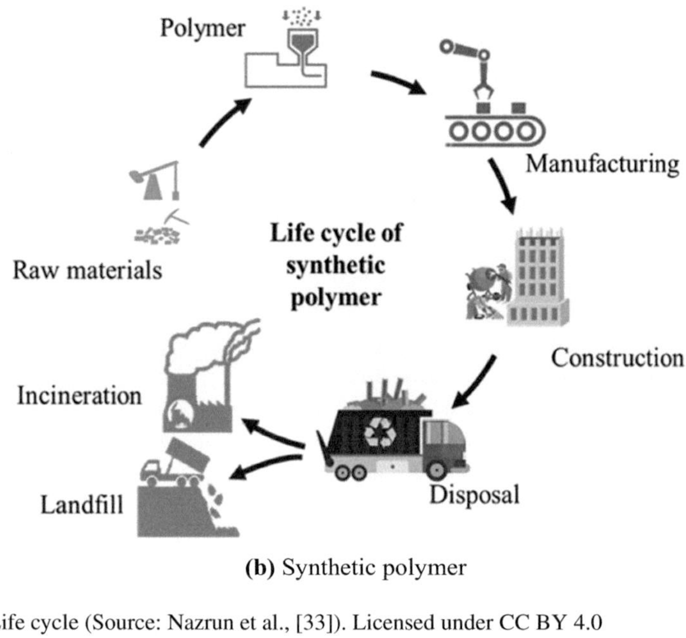

(b) Synthetic polymer

Fig. 8.5 Life cycle (Source: Nazrun et al., [33]). Licensed under CC BY 4.0

oil resins [38], cardanol resins [39], soy protein resins [40], wheat gluten protein resins [41], etc.

As shown in Refs. [42–44], these studies discuss a number of topics concerning the use of green composites in structures with a focus on their mechanical properties. For example, the prospects of natural fibre-reinforced green composites for

(a) Natural

(b) Synthetic

Fig. 8.6 Fibres examples [34]

secondary structural applications have been demonstrated in the study by Kopparthy et al. [28]. Nevertheless, articles on the use of entire green composites, which are composed of natural fibres and green resins, in structures are very scarce. This lack of comprehensive study contributes to a poor understanding of their performance, thereby limiting their practical implementation.

As a result, at present, polymer recycling could potentially serve as a viable pathway towards establishing a circular economy and achieve sustainable development. Despite efforts made by suppliers, the materials derived from discarded plastics or plastic debris streams remain a blend of various polymers, including polyethylene (PE), polypropylene (PP) and polyethylene-terephthalate (PET) [45–47]. The incompatibility between PP and PET has been reported by a number of research [48, 49]. Mechanical characteristics including strain at break, impact strength and the transition from ductile-to-brittle behaviour are all negatively affected by this incompatibility. Hence, understanding how the behaviour of polymer changes in relation to changes in its constituents and their proportions is crucial.

8.3 Homogenization Theories and Multi-scale Approaches

Composite material offers several advantages over conventional materials, with one of the main benefits is high strength-to-weight ratio. Because of this, composite materials have been employed in engineering applications for several decades and have become the main material for structural components in several industrial sectors, including aerospace, automotive and aircraft. Wood and human bones are two examples of composite materials that are found in nature. Composite materials feature complex microstructures that are optimized to sustain functional loads. Given their widespread use in nature, composite materials appear to offer several advantages and may soon be extensively utilized for a variety of structural applications. A simple, effective, precise and feasible approach to analysis is required due to the growing demand of composite materials.

Two separate techniques that are commonly combined in numerical frameworks for the analysis of composite materials are micro-mechanical analysis and macro-mechanical analysis. The composite structure is assumed as a homogeneous orthotropic continuum from a macro-mechanical perspective [50]. For the majority of technical applications, the micro-heterogeneity of the composite material is neglected and not taken into account, which is generally an acceptable consideration. However, more complex situations where the microscopic phenomenon greatly influences the material behaviour at the macro-scale—such as fibre buckling to microcracking—are better suited for the micro-mechanical analysis [51]. Hence, many multi-scale approaches that consider linked analysis at two separate scales, the macro-scales and micro-scales, are commonly found in literature.

Composites can be defined as heterogeneous materials composed of a minimum of two separate components. Composites commonly include partial, short and/or long fibres integrated in the matrix as reinforcing material. The composite materials exhibit several failure modes, most of which initiates in the microstructure and spread resulting in ultimate failure of the material. Thus, the material's microstructure needs to be incorporated into its constitutive model to project the failure of the composite material. Plenty of methods have been proposed to assess the behaviour of the composites which consider internal configuration of composite. The three most frequently used methods are homogenization, mixing theory and micro-models.

Since the micro-model can precisely model the internal structure, it is the most efficient method for modelling the microstructure of the composite. In addition to studying local effects like stress, strain and contact created by sliding steel balls on the composite surface [52, 53], the micro-model can be used to analyse matrix crack propagation from the debonded fibre [54], as well as to examine the impact of the interface on the macro tensile stress/strain curve of a particle-reinforced metal matrix composite [55], etc. The primary disadvantage of these models is their limited applicability due to their high computational cost.

Toupin and Truesdell [56] proposed the mixing theory. Adkins and Green later presented a general nonlinear equation [57]. Popov and Ortiz put forth a constitutive equation for unreinforced concrete, assuming it to be a composite material [58, 59]. Later on, Neamtu et al. [60] and Oller et al. [61] brought the series–parallel notion to the theory. The composite's mechanical properties are evaluated by examining the attributes of each constituent while also considering its topological distribution. Oller extended the mixing theory, and with it, any reinforced matrix composite can be analysed without being limited by the compatibility equation [62]. Rastellini et al. [63] proposed SP (series–parallel) continuum approach, according to which the composite's components will act as parallel materials in fibre direction and as series materials perpendicular to the direction of fibres. In order to investigate matrices reinforced with nanotubes, Otero et al. [64] updated the mixing theory.

The problem is broken down into two or more scales according to homogenization theory. One of the most relevant methods was developed and presented by Bensoussan et al. [65] and Sanchez-Palencia [66], and it is expressed in terms of asymptotic expansion theory. On the other hand, Suquet [67, 68] used the method of averages to answer and expand the two-scale problem to the nonlinear case. The fundamentals for first-order homogenization were laid by the Suquet [69]. To solve the micro-scale problem, the deformation gradient tensor is employed, and then the averaging equation is utilized to compute the macroscopic stress tensor. Moulinec [70], Terada [71, 72] and Miehe [73–75] have all contributed to the extension and application of the approach to massive deformations involving complex nonlinear material responses. Geers and Kouznetsova developed the second-order homogenization method [76–78]. This approach employs the deformation gradient tensor and its Lagrangian gradient to deal with the boundary value problem (B.V.P.) at the micro-scale. The second-order homogenization technique enables the solution of the macro-scale problem, even with moderate localization, by transferring macroscopic deformation gradients to the RVE borders. To address the issues of damage and fracture that lead to high degree of localization, a continuous-discontinuous homogenization is designed [79]. In order to adequately address the localization at a macroscopic scale, the homogenization can also be extended to a coupled thermomechanical issue [80] and cohesive zone problems [81]. Figure 8.7 provides an illustration of the fundamental idea of homogenization.

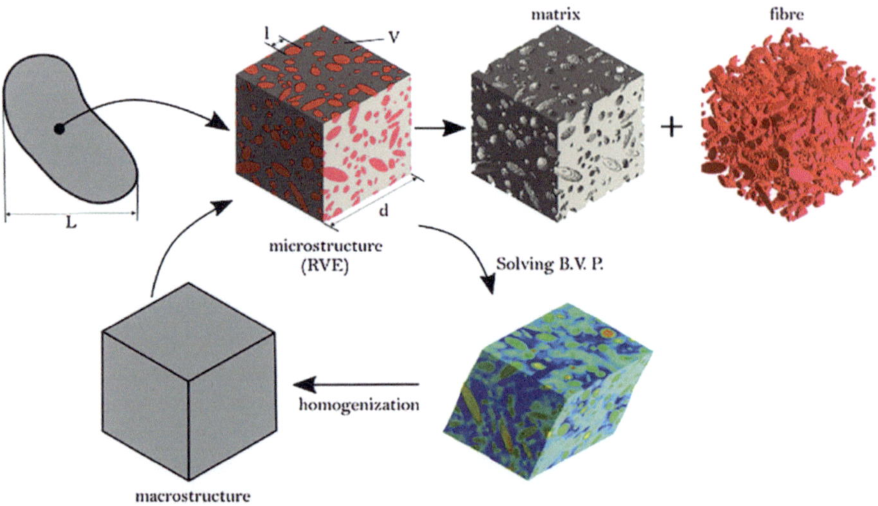

Fig. 8.7 Concept of homogenization (Source: Pesante et al., [82]). Licensed under CC BY 4.0.

References

1. Lacy P, Waughray D (2018) Harnessing the fourth industrial revolution for the circular economy: consumer electronics and plastics packaging. In: World economic forum
2. Galli A, Iha K, Moreno Pires S, Mancini MS, Alves A, Zokai G, Lin D, Murthy A, Wackernagel M (2020) Assessing the ecological footprint and biocapacity of Portuguese cities: critical results for environmental awareness and local management. Cities 96:102442
3. Stahel WR (2016) The circular economy. Nature 531:435–438
4. Geissdoerfer M, Savaget P, Bocken NMP, Hultink EJ (2017) The circular economy – a new sustainability paradigm? J Clean Prod 143:757–768
5. Within G (2015) A circular economy vision for a competitive Europe. McKinsey Center for Business and Environment
6. MacArthur E (2013) Others: towards the circular economy. J Ind Ecol 2(1):23–44
7. Parliament E (2023) Circular economy: definition, importance and benefits. News European Parliament, pp 23–26
8. Turner RK, Pearce DW (1990) The ethical foundations of sustainable economic development. International Institute for Environment and Development London, UK
9. Kirchherr J, Reike D, Hekkert M (2017) Conceptualizing the circular economy: an analysis of 114 definitions. Resour Conserv Recycl 127:221–232
10. Mhatre P, Panchal R, Singh A, Bibyan S (2021) A systematic literature review on the circular economy initiatives in the European Union. Sustain Prod Consumpt 26:187–202
11. Gehin A, Zwolinski P, Brissaud D (2008) A tool to implement sustainable end-of-life strategies in the product development phase. J Clean Prod 16(5):566–576
12. Hatcher GD, Ijomah WL, Windmill JFC (2011) Design for remanufacture: a literature review and future research needs. J Clean Prod 19(17–18):2004–2014
13. Ali MSM, Yusof RNR (2018) Others: intention to practice reduce, reuse & recycle (3R) among expatriates working in Malaysia. Int J Acad Res Bus Soc Sci 8(3):276–295
14. Bovea MD, Pérez-Belis V (2018) Identifying design guidelines to meet the circular economy principles: a case study on electric and electronic equipment. J Environ Manag 228:483–494
15. Arruda EH, Melatto RAPB, Levy W, de Melo Conti D (2021) Circular economy: a brief literature review (2015–2020). Sustain Operat Comput 2:79–86

16. Geng Y, Fu J, Sarkis J, Xue B (2012) Towards a national circular economy indicator system in China: an evaluation and critical analysis. J Clean Prod 23(1):216–224
17. (UNEP) UNEP (2019) Innovative pathways to achieve sustainable consumption and production. United Nations Environment Assembly
18. Macarthur E, Heading H et al (2019) Completing the picture: how the circular economy tackles climate change. Ellen MacArthur Found
19. Panel UIR (2018) The weight of cities: resource requirements of future urbanization. United Nations Environment Programme
20. Panel UIR (2018) Resource efficiency and climate change: material efficiency strategies for a low-carbon future. United Nations Environment Programme
21. Marichova A (2021) Circular economy and strategic management of the construction company. Ovidius University Ann Const Ser Civil Eng 23(1):35–44
22. Park S-J, Seo M-K (2011) Types of composites. In: Park S-J, Seo M-K (eds) Interface science and composites. Elsevier, pp 501–629
23. Gavalda Diaz O, Garcia Luna G, Liao Z, Axinte D (2019) The new challenges of machining ceramic matrix composites (CMCs): review of surface integrity. Int J Mach Tools Manuf 139:24–36
24. Mouritz AP (2012) 16 - Metal matrix, fibre–metal and ceramic matrix composites for aerospace applications. In: Introduction to aerospace materials. Woodhead Publishing, pp 394–410
25. Matias ÁA, Lima MS, Pereira J, Pereira P, Barros R, Coelho JFJ, Serra AC (2020) Use of recycled polypropylene/poly(ethylene terephthalate) blends to manufacture water pipes: an industrial scale study. Waste Manag 101:250–258
26. Wright L (2019) Plastic warms the planet twice as much as aviation–here's how to make it climate-friendly
27. Dickinson TIM (2020) How big oil and big soda kept the plastics crisis a secret for decades - rolling stone
28. Kopparthy SDS, Netravali AN (2021) Review: green composites for structural applications. Compos Part C: Open Access 6:100169
29. Gómez EF, Michel FC Jr (2013) Biodegradability of conventional and bio-based plastics and natural fiber composites during composting, anaerobic digestion and long-term soil incubation. Polym Degrad Stab 98(12):2583–2591
30. Jmal H, Bahlouli N, Wagner-Kocher C, Leray D, Ruch F, Munsch J-N, Nardin M (2018) Influence of the grade on the variability of the mechanical properties of polypropylene waste. Waste Manag 75:160–173
31. Pessey ND, Bahlouli SP, Ahzi S (2008) Polymer composites for the automotive industry: characterisation of the recycling effect on the strain rate sensitivity. Int J Crashworthiness 13(4):411–424
32. Shanmugam V, Mensah RA, Försth M, Sas G, Restás Á, Addy C, Xu Q, Jiang L, Neisiany RE, Singha S (2021) Others: circular economy in biocomposite development: state-of-the-art, challenges and emerging trends. Compos Part C: Open Access 5:100138
33. Nazrun T, Hassan MK, Hossain MD, Ahmed B, Hasnat MR, Saha S (2024) Application of biopolymers as sustainable cladding materials: a review. Sustainability 16(1)
34. Hasan Shuvo Z, Rahman MZ, Saha B, Hossain Ador MS (2024) 13.16 - Composite materials from synthetic and natural sources: fabrication techniques and applications. In: Hashmi S (ed) Comprehensive materials processing, 2nd ed. Elsevier, Oxford, pp 303–328
35. Dubrovski PD (2010) Woven fabric engineering. IntechOpen
36. Cigasova J, Stevulova N, Sicakova A, Junak J (2013) Some aspects of lightweight composites durability. Chem Eng Trans 32:1615–1620
37. Varghese S, Karger-Kocsis J, Gatos KG (2003) Melt compounded epoxidized natural rubber/layered silicate nanocomposites: structure-properties relationships. Polymer 44(14):3977–3983
38. Mosiewicki M, Borrajo J, Aranguren MI (2005) Mechanical properties of woodflour/linseed oil resin composites. Polym Int 54(5):829–836

39. Aggarwal LK, Thapliyal PC, Karade SR (2007) Anticorrosive properties of the epoxy–cardanol resin based paints. Prog Org Coat 59(1):76–80
40. Kim JT, Netravali AN (2010) Effect of protein content in soy protein resins on their interfacial shear strength with ramie fibers. J Adhes Sci Technol 24(1):203–215
41. Kim NK, Bruna FG, Das O, Hedenqvist MS, Bhattacharyya D (2020) Fire-retardancy and mechanical performance of protein-based natural fibre-biopolymer composites. Compos Part C: Open Access 1:100011
42. Dicker MPM, Duckworth PF, Baker AB, Francois G, Hazzard MK, Weaver PM (2014) Green composites: a review of material attributes and complementary applications. Compos A Appl Sci Manuf 56:280–289
43. Koronis G, Silva A, Fontul M (2013) Green composites: a review of adequate materials for automotive applications. Compos B Eng 44(1):120–127
44. Saba N, Jawaid M, Sultan MTH, Alothman OY (2017) Green biocomposites for structural applications. Green biocompos Des Appl 1–27
45. Serranti S, Luciani V, Bonifazi G, Hu B, Rem PC (2015) An innovative recycling process to obtain pure polyethylene and polypropylene from household waste. Waste Manag 35:12–20
46. Wang C-Q, Wang H, Liu Y-N (2015) Separation of polyethylene terephthalate from municipal waste plastics by froth flotation for recycling industry. Waste Manag 35:42–47
47. Tang P, Brouwers HJH (2017) Integral recycling of municipal solid waste incineration (MSWI) bottom ash fines (0–2 mm) and industrial powder wastes by cold-bonding pelletization. Waste Manag 62:125–138
48. Teh JW, Rudin A, Keung JC (1994) A review of polyethylene–polypropylene blends and their compatibilization. Adv Polym Technol 13(1):1–23
49. Bertin S, Robin J-J (2002) Study and characterization of virgin and recycled LDPE/PP blends. Eur Polym J 38(11):2255–2264
50. Devireddy SBR, Biswas S (2014) Effect of fiber geometry and representative volume element on elastic and thermal properties of unidirectional fiber-reinforced composites. J Compos 2014
51. de Carvalho MV (2016) Computational framework for multi-scale analysis of heterogeneous materials using digital image recognition. University of Porto
52. Váradi K, Néder Z, Friedrich K (2004) FE contact and thermal analysis of composite surfaces with different fibre orientations subjected to a sliding steel asperity. Finite Elem Anal Des 40(11):1475–1497
53. Goda T, Váradi K, Friedrich K (2001) Fe micro-models to study contact states, stresses and failure mechanisms in a polymer composite subjected to a sliding steel asperity. Wear 251(1):1584–1590
54. Huang H, Talreja R (2006) Numerical simulation of matrix micro-cracking in short fiber reinforced polymer composites: initiation and propagation. Compos Sci Technol 66(15):2743–2757
55. Yuan MN, Yang YQ, Li C, Heng PY, Li LZ (2012) Numerical analysis of the stress–strain distributions in the particle reinforced metal matrix composite SiC/6064Al. Mater Des 38:1–6
56. Truesdell C, Toupin R (1960) The classical field theories. Springer
57. Green AE, Adkins JE (1964) A contribution to the theory of non-linear diffusion. Arch Ration Mech Anal 15(3):235–246
58. Ortiz M, Popov EP (1982) A physical model for the inelasticity of concrete. Proc Royal Soc Lond A Math Phys Sci 383(1784):101–125
59. Ortiz M, Popov EP (1982) Plain concrete as a composite material. Mech Mater 1(2):139–150
60. Neamtu L, Oller S, Oñate E. A generalized mixing theory elasto-damage-plastic model for finite element analysis of composites. In: Computational plasticity: fundamentals and applications, Pts 1 and 2, pp 1214–1219 (1997).
61. Oller S, Neamtu L, Oñate E (1995) Una generalización de la teor\'\ia de mezclas clásica para el tratamiento de compuestos en serie/paralelo. In: Congreso Nacional de Materiales Compuestos pp 433–438
62. Oller S (2003) Simulación numérica del comportamiento mecánico de los materiales compuestos. Centro Internacional de Métodos Numéricos en Ingenier\'\ia

63. Rastellini F, Oller S, Salomón O, Oñate E (2008) Composite materials non-linear modelling for long fibre-reinforced laminates: continuum basis, computational aspects and validations. Comput Struct 86(9):879–896
64. Otero F, Martínez X, Oller S, Salomón O (2012) Study and prediction of the mechanical performance of a nanotube-reinforced composite. Compos Struct 94(9):2920–2930
65. Bensoussan A, Lions J-L, Papanicolaou G (2011) Asymptotic analysis for periodic structures. Am Math Soc
66. Sánchez-Palencia E (1980) Non-homogeneous media and vibration theory. Lecture Note in Physics, Springer-Verlag 320:57–65
67. Suquet P (1982) Plasticite et homogenesation. Université Pierre et Marie Curie
68. Suquet P (1987) Elements of homogenization for inelastic solid mechanics. In: Homogenization techniques for composite media
69. Suquet PM (1985) Local and global aspects in the mathematical theory of plasticity. In: Sawczuk A et al (eds) Plasticity Today
70. Moulinec H, Suquet P (1998) A numerical method for computing the overall response of nonlinear composites with complex microstructure. Comput Methods Appl Mech Eng 157(1):69–94
71. Terada K, Hori M, Kyoya T, Kikuchi N (2000) Simulation of the multi-scale convergence in computational homogenization approaches. Int J Solids Struct 37(16):2285–2311
72. Terada K, Kikuchi N (2001) A class of general algorithms for multi-scale analyses of heterogeneous media. Comput Methods Appl Mech Eng 190(40–41):5427–5464
73. Miehe C, Schotte J, Schröder J (1999) Computational micro–macro transitions and overall moduli in the analysis of polycrystals at large strains. Comput Mater Sci 16(1):372–382
74. Miehe C, Koch A (2002) Computational micro-to-macro transitions of discretized microstructures undergoing small strains. Arch Appl Mech 72(4–5):300–317
75. Miehe C, Schotte J, Lambrecht M (2002) Homogenization of inelastic solid materials at finite strains based on incremental minimization principles. Application to the texture analysis of polycrystals. J Mech Phys Solids 50(10):2123–2167
76. Geers MGD, Kouznetsova V, Brekelmans WAM (2001) Meso-macro scale transitions-2 Strain gradients and higher-order continua-Gradient-enhanced computational homogenization for the micro-macro scale transition. J Phys Colloq 11(5):145–152
77. Kouznetsova V, Geers MGD, Brekelmans WAM (2002) Multi-scale constitutive modelling of heterogeneous materials with a gradient-enhanced computational homogenization scheme. Int J Numer Meth Eng 54(8):1235–1260
78. Kouznetsova VG, Geers MGD, Brekelmans WAM (2004) Multi-scale second-order computational homogenization of multi-phase materials: a nested finite element solution strategy. Comput Methods Appl Mech Eng 193(48–51):5525–5550
79. Massart TJ, Peerlings RHJ, Geers MGD (2007) Structural damage analysis of masonry walls using computational homogenization. Int J Damage Mech 16(2):199–226
80. Özdemir I, Brekelmans WAM, Geers MGD (2008) FE2 computational homogenization for the thermo-mechanical analysis of heterogeneous solids. Comput Methods Appl Mech Eng 198(3):602–613
81. Matouš K, Kulkarni MG, Geubelle PH (2008) Multiscale cohesive failure modeling of heterogeneous adhesives. J Mech Phys Solids 56(4):1511–1533
82. Pesante KP, Saavedra GP (2021) Mechanical properties of a wood flour-PET composite through computational homogenisation. Comput Mater Continua 67(3):4061–4079
83. ADITYA VIDWANS. Industrial or urban recyclable materials for structural applications: green composites with natural fillers and metals. PhD thesis, Sapienza University of Rome, 2025. https://hdl.handle.net/11573/1745888

MIX
Papier aus verantwortungsvollen Quellen
Paper from responsible sources
FSC® C105338

If you have any concerns about our products,
you can contact us on
ProductSafety@springernature.com

In case Publisher is established outside the EU,
the EU authorized representative is:
Springer Nature Customer Service Center GmbH
Europaplatz 3, 69115 Heidelberg, Germany

Printed by Libri Plureos GmbH
in Hamburg, Germany